BARRON'S NEW JERSEY GRADE 3 MATH TEST

Thomas Walsh, M.A., M.Ed, Ed.D.
Associate Professor, Math and Science Education

Dan Nale, MBA, M.Ed.
Supervisor of Math, Science, and Technology, K-5

Acknowledgments

For her understanding, support, and forbearance during my absences rewriting this book, I would like to thank my wife Elaine. I would also like to thank everyone at Kean University (particularly Dean Susan Polirstok and President Dawood Farahi) who encouraged me to finish this project.

Thomas Walsh

For understanding my long hours at the computer, I'd like to thank my wife Teresa and our two daughters, Gianna and Victoria. I am looking forward to spending more family time together. Endless gratitude goes to my parents Joel and Michelle for instilling in me values of perseverance and determination. I would also like to thank the Pemberton Township School District for giving me the opportunity to make a more global impact through a focus on mathematics education.

Dan Nale

All inquiries should be addressed to:
Barron's Educational Series, Inc.
250 Wireless Boulevard
Hauppauge, New York 11788
www.barronseduc.com

ISBN-978-1-4380-0562-1
Library of Congress Control Number: 2014953553

Date of Manufacture: December 2014
Manufactured by: B11R11

Printed in the United States of America
9 8 7 6 5 4 3 2 1

10%
POST-CONSUMER WASTE
Paper contains a minimum of 10% post-consumer waste (PCW). Paper used in this book was derived from certified, sustainable forestlands.

Contents

IMPORTANT NOTE: Barron's has made every effort to ensure the content of this book is accurate as of press time, but the PARCC Assessments are constantly changing. Be sure to consult *https://www.parcconline.org/* for all the latest testing information. Regardless of the changes that may be announced after press time, this book will still provide a strong framework for third-grade students preparing for the assessment.

Introduction to the New Jersey Math Test for Parents and Teachers

CHAPTER 1

Overview of this Book

This chapter provides you with a clear picture of the New Jersey State educational assessment: The Partnership for Assessment of Readiness for College and Careers (PARCC) and the Grade 3 Common Core State Standards (CCSS). Along with the content standards, this book also contains Standards for Mathematical Practice (SMP) that span multiple grades. Chapter 2 delves into these standards and provides strategies for working with Grade 3 concepts. After that, you will find a chapter dedicated to each domain of the CCSS along with a practice PARCC Performance-Based Assessment (PBA) and an End-of-Year Assessment (EOY). Additionally, in Appendixes A and B you will find the CCSS and the SMP, respectively, in their full-text form.

Common Core Overview

In June 2010, the New Jersey State Board of Education (NJBOE) and the New Jersey Department of Education (NJDOE) adopted the CCSS. These standards were developed in collaboration with teachers and school administrators to provide a clear and consistent framework to prepare children for college and the workforce. (Source: *http://www.state.nj.us/education/sca/*)

The state-led effort to develop the CCSS was launched in 2009 by state leaders, including governors and state commissioners of education from 48 states, two territories, and the District of Columbia, through their membership in the National Governors Association Center for Best Practices (NGA Center) and the Council of Chief State School Officers (CCSSO). State school chiefs and governors recognized the value of consistent, real-world learning goals. They launched this effort to ensure that all students, regardless of where they live, graduate from high school prepared for college, career, and life. (Source: *http://www.corestandards.org/about-the-standards/development-process/*)

The standards were based on the following:

- The best state standards already in existence
- The experience of teachers, content experts, and state departments of education
- Feedback from the public

Common Core Introduction

In Grade 3, instructional time should focus on four critical areas: (1) developing an understanding of multiplication and division and of strategies for multiplication and division within 100, (2) developing an understanding of fractions, especially unit fractions (fractions with a numerator of 1), (3) developing an understanding of the structure of rectangular arrays and of area, and (4) describing and analyzing two-dimensional shapes.

1. Students develop an understanding of multiplication and division of whole numbers through activities and problems involving equal-sized groups, arrays, and area models. Multiplication is finding an unknown product, and division is finding an unknown factor. For equal-sized groups, division can require finding the unknown number of groups or the unknown group size. Students use properties of operations to calculate products of whole numbers, using increasingly sophisticated strategies based on these properties to solve multiplication and division problems involving single-digit factors. By comparing a variety of solution strategies, students learn the relationship between multiplication and division.

2. Students develop an understanding of fractions, beginning with unit fractions. They view fractions in general as being built out of unit fractions. Students use fractions along with visual fraction models to represent parts of a whole. Students understand that the size of a fractional part is relative to the size of the whole. For example, 1/2 of the paint in a small bucket could be less paint than 1/3 of the paint in a larger bucket. However, 1/3 of a ribbon is longer than 1/5 of the same ribbon. Students are able to use fractions to represent numbers equal to, less than, and greater than 1. They solve problems that involve comparing fractions by using visual fraction models and strategies based on noticing equal numerators or denominators.

3. Students recognize area as an attribute of two-dimensional regions. They measure the area of a shape by finding the total number of same-size units of area required to cover the shape without gaps or overlaps, a square with sides of unit length being the standard unit for measuring area. Students understand that rectangular arrays can be decomposed into identical rows or into identical columns. By decomposing rectangles into rectangular arrays of squares, students connect area to multiplication and justify using multiplication to determine the area of a rectangle.

4. Students describe, analyze, and compare properties of two-dimensional shapes. They compare and classify shapes by sides and angles. Students connect these attributes with definitions of shapes. Students also relate fraction work to geometry by expressing the area of part of a shape as a unit fraction of the whole.

Mathematical Content

The New Jersey PARCC 3 Math Exam covers two broad sets of standards. The first set, the mathematical content, is outlined in this section.

Grade 3 Common Core State Standards Overview

Operations and Algebraic Thinking	
	• Represent and solve problems involving multiplication and division. • Understand properties of multiplication and the relationship between multiplication and division. • Multiply and divide within 100. • Solve problems involving the four operations, and identify and explain patterns in arithmetic.
Number and Operations in Base Ten	
	• Use place value and properties of operations to perform multidigit arithmetic.
Number and Operations—Fractions	
	• Develop an understanding of fractions as numbers.

Measurement and Data	
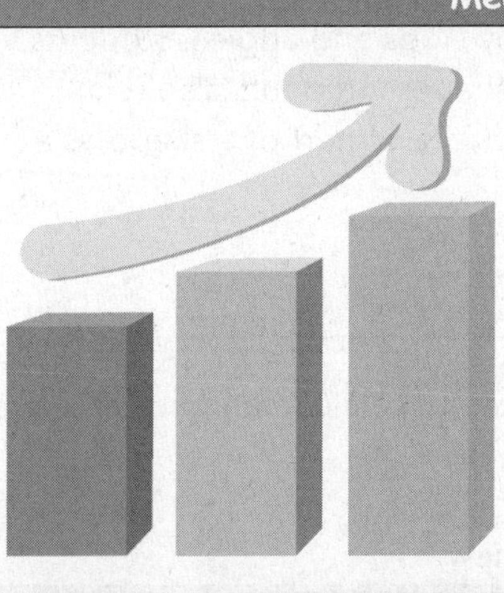	• Solve problems involving measurement and estimation of intervals of time, liquid volumes, and masses of objects. • Represent and interpret data. • Geometric measurement: understand concepts of area and relate area to multiplication and to addition. • Geometric measurement: recognize perimeter as an attribute of plane figures and distinguish between linear and area measures.
Geometry	
	• Reason with shapes and their attributes.

Standards for Mathematical Practice

The second set of standards covered by the New Jersey PARCC 3 Math Exam is the mathematical practice (SMP). These standards are used throughout grades K-12. The SMP standards are outlined in this section.

Standards for Mathematical Practices

MP1		• Make sense of problems, and keep trying to solve. Understand a problem, and work on it until you solve it. Do not give up on the problem until you have solved it.

MP2		• Reason abstractly and quantitatively. Make mathematical assumptions, and evaluate them to solve the problem at hand. Devise an equation to express the problem.
MP3		• Construct viable arguments, and critique the reasoning of others. Organize the ideas for solving a problem, and communicate them mathematically. Constructively judge your ideas and those of others to solve the problem.
MP4		• Model with mathematics. Construct and use mathematical models or representations of a problem to help you make sense of the problem and solve it.
MP5		• Use appropriate tools strategically. Appropriate tools could include paper and pencil, a calculator, a manipulative model, a computer, a protractor, a ruler, play money, or other. Just remember to use the right tool to solve the specific problem.
MP6	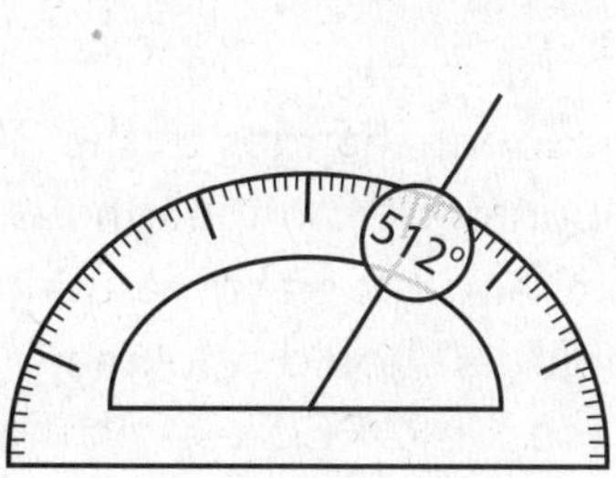	• Be precise. Use clear definitions, and use mathematical symbols correctly to solve the problem. State the proper units of measure, and use the right amount of accuracy that the problem requires. For example, don't use inches to explain a problem about the number of miles.

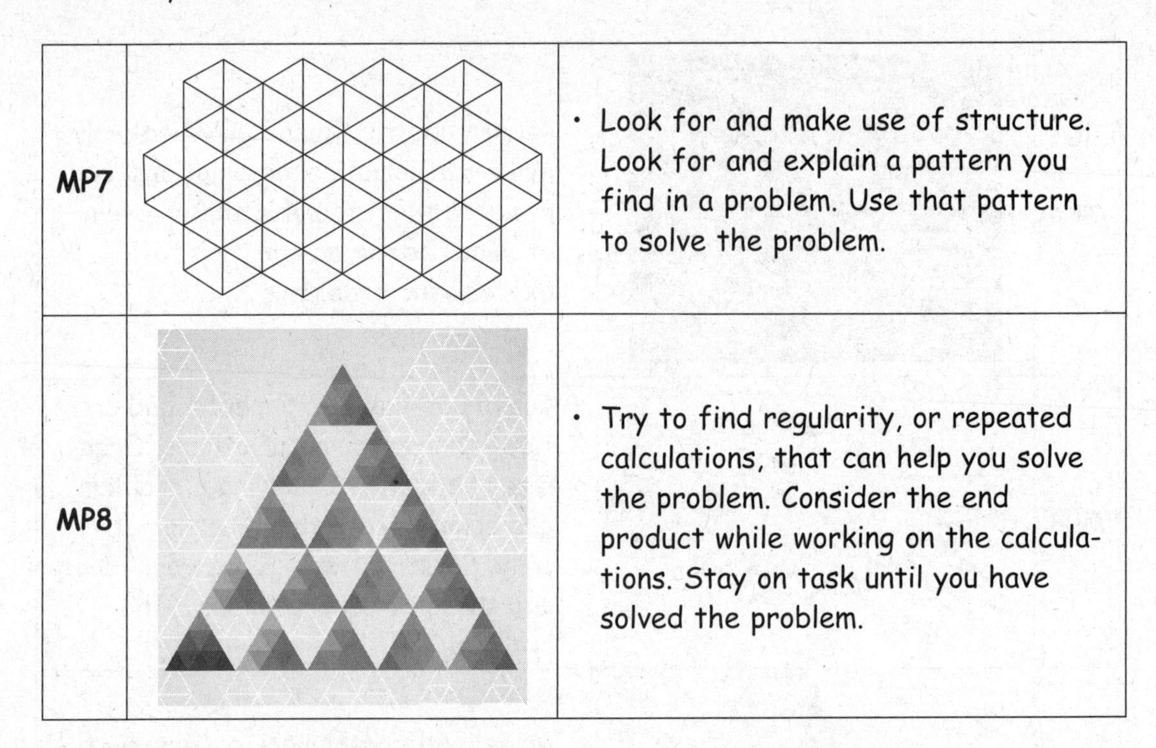

MP7		• Look for and make use of structure. Look for and explain a pattern you find in a problem. Use that pattern to solve the problem.
MP8		• Try to find regularity, or repeated calculations, that can help you solve the problem. Consider the end product while working on the calculations. Stay on task until you have solved the problem.

In Chapter 2, we will use the SMP standards to solve some problems.

Introduction to the PARCC

The PARCC assessment was developed as a way to address the CCSS. Past assessments, including the NJ ASK, were not equipped to handle the rigor and depth of the CCSS. Because the CCSS emphasizes strategies and explanations and also focuses on a few concepts, assessments focused solely on multiple-choice questions are no longer acceptable. The PARCC assessment moves away from the multiple-choice tests of the past and instead uses both short and extended constructed-response questions. In short-response questions, students simply input an answer, such as a number. In extended-response questions, students input the answer and then explain the processes and strategies they used to find that answer. In addition, students are required to use online manipulatives to demonstrate understanding of conepts.

Parents and educators should stay on top of current changes to the PARCC. All information contained in this chapter has been sourced from the official PARCC website. However, this information is subject to change at any time during/after production of this book. For this reason, stakeholders should bookmark the following website and use it as a primary resource for PARCC test planning: *https://www.parcconline.org/*.

As of the publication date of this book, the mathematics portion of the PARCC contains two required components: a performance-based assessment (PBA) and an end-of-year assessment (EOY). The assessments are then broken down to contain a set amount of Type I, Type II, and Type III questions. As per the PARCC High-Level Blueprint:

- Type I: Tasks assessing concepts, skills, and procedures
- Type II: Tasks assessing expressing mathematical reasoning
- Type III: Tasks assessing modeling/applications

The PBA is administered after approximately 75% of the school year. The mathematics PBA at each grade level includes both short- and extended-response questions focused on conceptual knowledge and skills and the SMP for reasoning and modeling. This assessment is administered in two sessions, with each session taking approximately 50 minutes. As per the PARCC High-Level Blueprint, the Grade 3 PBA is comprised of the following:

- 8 Type I 1-pt questions
- 2 Type I 2-pt questions
- 2 Type II 3-pt questions
- 2 Type II 4-pt questions
- 2 Type III 3-pt questions
- 1 Type III 6-pt question

The EOY assessment is administered after approximately 90% of the school year. The mathematics EOY assessments are comprised primarily of short-answer questions focused on conceptual knowledge, skills, and understandings. This assessment is administered in two sessions, with each session taking approximately 50 minutes. As per the PARCC High-Level Blueprint, the Grade 3 PBA is comprised of the following:

- 34 Type I 1-pt questions
- 5 Type I 2-pt questions

Although estimated times are given for each session, students have a set amount of additional time for each session to provide them with ample time to demonstrate their knowledge.

PARCC Expectations

All students taking the PARCC assessment need to be familiar with technology and with inputting responses into a computer. Unlike the previous state assessment (NJ ASK), the PARCC is computerized. This means students take the entire assessment on a personal computer. Because most districts do not have a 1 : 1 ratio of computers to students, the assessment will likely be administered in staggered schedules where groups of students will complete the assessment at a time.

To use their time most efficiently while taking the assessment, students should be familiar with using a mouse and keyboard. On the keyboard, students may need to input letters, numbers, and other characters as well, such as <, >, =, and so on. Students should be familiar with how to enter these mathematical symbols and characters in order to decrease frustration levels and to use their test-taking time efficiently. Similarly, students should be able to use a mouse or touchpad to select these characters from the screen in order to input their response.

Students also need to be comfortable with sitting at a computer and staring at a screen for long periods of time. Each session of the assessment will take an estimated 50+ minutes, if not longer. Because students will be looking at a computer screen for such an extended time, they should incorporate strategies to reduce eye strain, such as intermittently looking away from the screen and focusing on a distant object. Doing this will refocus the students' eyes and help reset thought processes when moving to a new section.

How the PARCC Is Given

The testing authority chosen for the State of New Jersey is PARCC. The assessments given by PARCC are administered on a computer. When a question is posed on the computer, there will be four or more choices for the answer. The student must hold the arrow on the correct choice and drag it to the answer blank (below the choices). In a number of questions, there will be two or more answers required, and there will be a corresponding number of answer blanks available for those choices. Some questions will require manipulating a pointer on a number line, or a small explanation, by the student. Some questions have two parts (Part A and Part B), requiring some extra calculation on the part of the student. Some questions will require completing a bar graph or a pie chart. Please look carefully at the examples that have been provided by PARCC (printed below in this book):

Type I Sample Question:

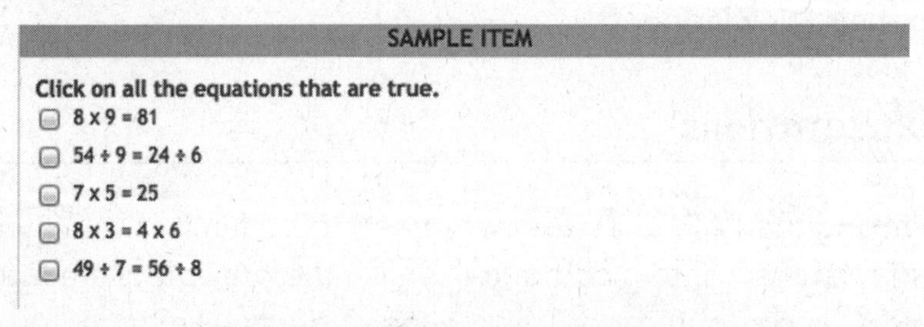

Type II Sample Question:

SAMPLE ITEM

Part A
A farmer plants 3/4 of the field with soybeans.
Drag the soybean to the field as many times as needed to show the fraction of the field that is planted with soybeans.

Farmer's Field

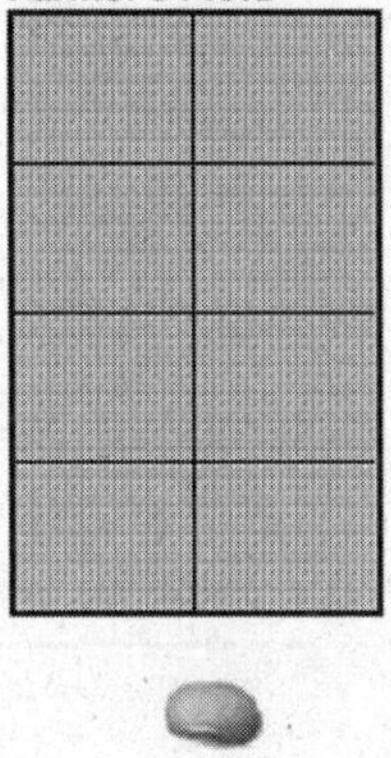

Soybean

Reset

SAMPLE ITEM

Part B
Type a fraction different than 3/4 in the boxes that also represents the fractional part of the farmer's field that is planted with soybeans.

$\frac{3}{4} = \frac{\square}{\square}$

Farmer's Field

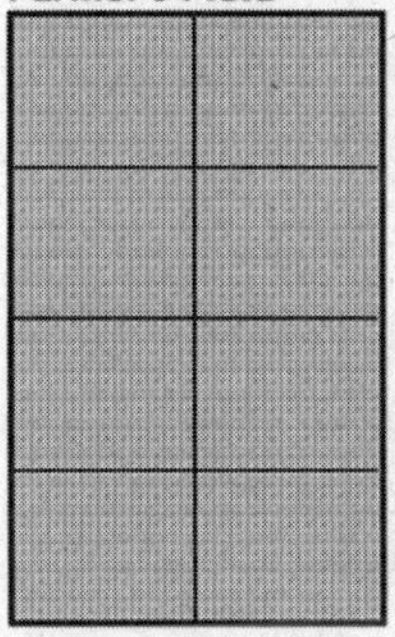

Reset

Explain why the two fractions above are equal.

Type III Sample Question:

An art teacher will tile a section of the wall with painted tiles made by students in three art classes.

- Class A made 18 tiles
- Class B made 14 tiles.
- Class C made 16 tiles.

Part A

What is the total number of tiles that are to be used?

[ ] tiles

Part B

The grid shows how much wall space the art teacher can use. Use the grid to create a rectangular array showing how the art teacher might arrange the tiles on the wall.

Select the boxes to shade them. Each tile should be shown by one shaded box.

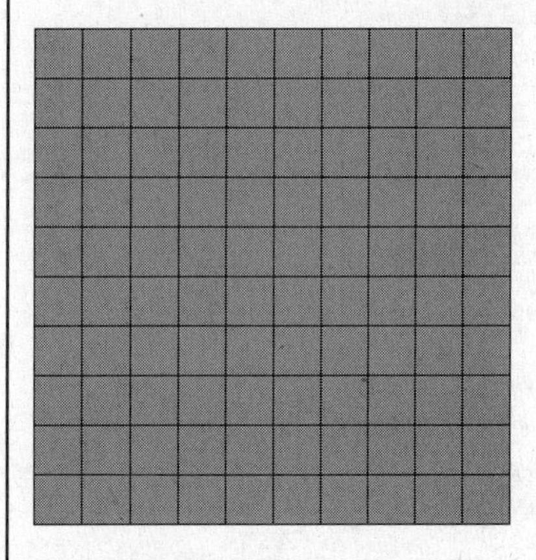

Part C

Andy created a rectangular array showing how he would place 56 small tiles on the wall. He placed 7 tiles in each row. He wrote a multiplication equation using *R* to stand for the number of rows he used.

Write an equation using *R* that Andy could have written.

CUT PASTE UNDO REDO

CHAPTER 2

Standards for Mathematical Practice

In this chapter, you will do some problems so you can see how to implement the Standards for Mathematical Practice (SMP). Keep the end product in mind; you want to solve a problem. You do not want to get lost in the standards and not solve the problem. Use the standards to help you solve the problem.

Problem Solving

All of the SMP are used in problem solving, and you solve problems every day. A problem is a situation you face where you don't have a clear set of steps to solve it or that you've never seen before. Facing a situation like that can be scary for many, but there are ways to approach it.

Keep in mind these points: You are not born with better or worse problem-solving ability than anyone else. Problem solving is a skill, not a talent. All skills are improved by practice, including problem solving. The best problem solvers get better with practice. Do not *ever* believe that you cannot solve problems, because you *can*. You just need to practice at it. You should remember a few things as you attempt to solve a problem.

1. Persevere in solving a problem. Believe that you can do it, because you *can*! If you believe you cannot solve the problem, you won't even try. If you believe you can, you'll try and try and try again until you succeed.
2. Don't think there's only one way to solve a particular problem. Very often there is more than one way to solve a particular problem. We'll look at a few ways later in this chapter.
3. Keep trying! You can solve it if you keep at it. If you give up, you certainly won't solve it.
4. Look for the pattern in all of these solutions. Remember that all mathematics is simply looking for patterns, making sure the pattern exists, and attempting to explain the pattern.

Strategies for Problem Solving

One of the most widely used methods for solving a problem has four steps:

1. Make sense of the problem–understand all aspects of the problem. Try to find out all the important information about the problem, and understand what is actually asked in the problem. What information given in the problem is needed to solve the problem? What information given in the problem is not needed to solve the problem?
2. Make a plan. Once you've understood all the aspects of the problem, you need to select the strategy to solve the problem. Could you draw a picture? Could you make a chart or table? Could you think of a simpler problem? Could you guess the answer and then check it? Could you work backward from the end to the beginning? We'll look at each of these later on.
3. Carry out the plan. Now draw the picture, make up the table, write down the simpler problem, or do the guessing and checking to find the answer to the problem.
4. Check your solution. Now that you have an answer, stop! Look at the answer, and think about it. Does it make sense? Does it answer the question completely? This last step is very important. However, many people do not do it and give the wrong answer. For example, suppose you had to find out the age of a person. You thought of a plan, carried out the plan, and figured that the person is 900 years old! That answer, of course, makes no sense. However, people new to problem solving will often give an answer that makes no sense just because they don't stop to think about the answer.

There are many strategies for solving problems. Here are six (lettered A through F).

A. Draw a Picture

If you are trying to tell someone how to get from your home to your school, you might draw a picture of the roads the person needs to take. Then you can list what roads the person should look for and how many turns he or she should make to get there. Very often a picture or a diagram of some kind will show the pattern clearly.

Example: Davy wanted to have an 8-foot-long pine board cut into 8 pieces. The lumber yard charges $6.00 to cut a board into 4 pieces, and each cut costs the same. How much will the lumber yard charge Davy?

We'll use the four-step process to solve this problem.

1. Make sense of the problem. We want to know how much Davy will be charged for cutting a board into 8 pieces. The useless information is the name Davy (it could be any person at all) and the fact that the wood is pine. We know how much it will cost for the lumber yard to cut the board into 4 pieces. How can we solve this problem?
2. Make a plan. Drawing a picture would make the number of cuts clearer. That seems to be a good way to move ahead. We'll draw a picture of the board and show the number of cuts. Then we can figure how much each cut will cost and finally how much cutting the board will cost overall.
3. Carry out the plan. The board cut into 4 pieces looks like this:

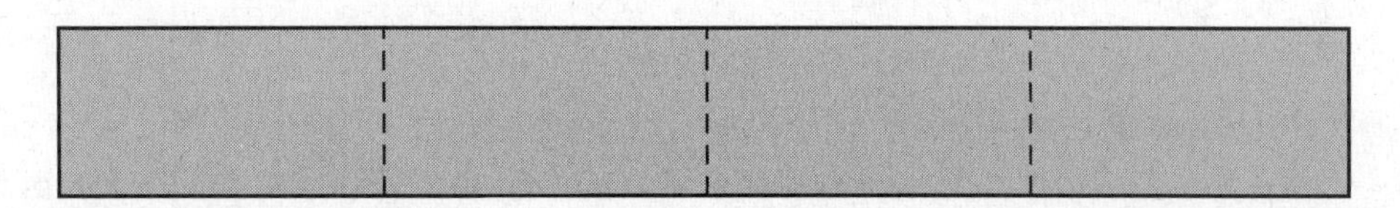

As we see, the board requires 3 cuts be made for the board to be in 4 pieces. Since the lumber yard charges $6.00 to make 3 cuts, it is charging $6/3 = $2 per cut. Now we must extend this to 8 cuts.

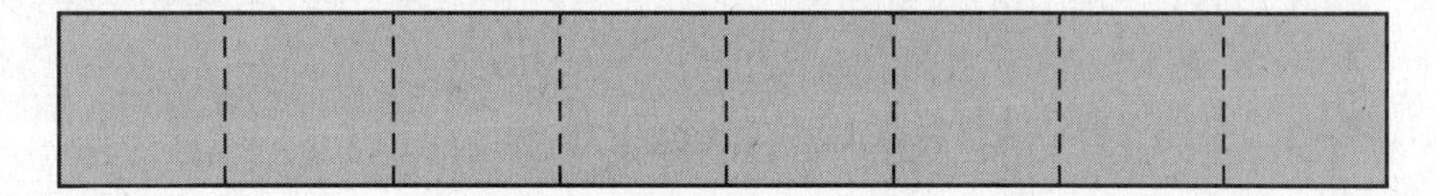

To cut the board into 8 pieces, we use seven cuts. Since each cut is $2, the lumber yard will charge $2 × 7 = $14.

4. Check your solution. Now that we have the answer, let's check to see if it satisfies all the conditions of the problem. Look at the last figure. It shows that 7 cuts gave us 8 pieces. Then 7 cuts times $2.00 per cut is $14.00. So the solution does satisfy the conditions of the problem.

Another Example: Mallory likes art and science, and she has several books on each subject. She has 14 books on art and 3 times as many books on science. How many books on science does she have? How many books on art and science does she have in all?

Use the four-step process:

1. Make sense of the problem. We want to find two values: the number of science books and the number of books on art and science altogether. As with the first example, the name of the person and the subjects of the books do not matter when solving the problem. How can we solve this?

2. Make a plan. Drawing a picture for this problem would work well. Since Mallory has 3 times the number of science books as art books, drawing a bar to show the number of art and science books can make the problem clearer.

3. Carry out the plan. Draw a bar to show the number of art books. Draw a second bar underneath the first to show the number of science books.

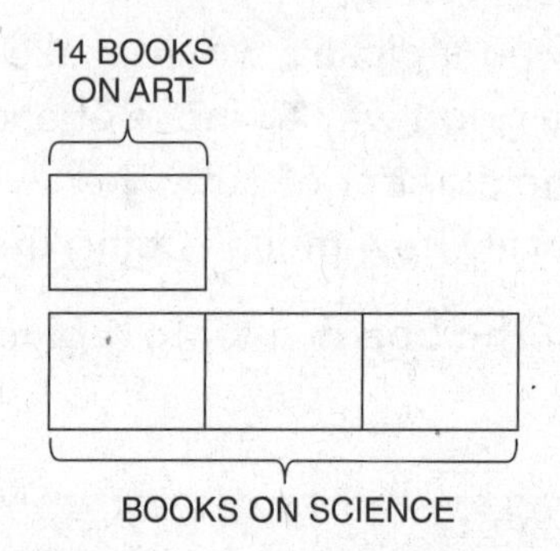

Now we see that the bar for the science books is 3 times the size of the bar for the art books. We know the number of art books (14). To solve, then, we will multiply 14 by 3 to get the number of science books: $14 \times 3 = 42$. The number of science books Mallory has is 42.

Finally, we find the number of books on art and science altogether: $14 + 42 = 56$. The total number of books on art and science that Mallory has is 56.

4. Check your solution. Do the answers make sense? To check, remember that there are 3 times as many science books as art books. To check, divide 42 by 3: $42 \div 3 = 14$. Since Mallory has 14 art books, this is correct. Then subtract $56 - 42 = 14$. This tells us that 56 is the total number of books on art and science that Mallory has. So both answers check out.

B. Make a Chart or a Table

You have a pocket full of change, and you want to know how much money you have. In this case, use a table to write down how many quarters, dimes, nickels, and pennies. Using a table will make it easy to tell you how much money you have. Sometimes a chart, table, or diagram of some kind will reveal the pattern to the solution.

Example: Grace had pennies, nickels, dimes, and quarters in the pocket of her jumpsuit. If she reached in and pulled out two coins, how many different combinations could she have?

Use the four-step process.

1. Make sense of the problem. We want to find out how many different combinations of two coins Grace can make with four different kinds of coins. The coins are pennies, nickels, dimes, and quarters. The useless information is that she's wearing a jumpsuit. How can we solve this?
2. Make a plan. If we made a table, we could put down all the combinations. That way, we would avoid duplicating combinations and make sure all combinations are used. We do not need to know how much each combination is worth in cents. The question doesn't ask for that. Using a table seems to be the best way to proceed.
3. Carry out the plan: The table looks like this:

   P–Penny; N–Nickel; D–Dime; Q–Quarter

Coins Used	Combination Number
P P	1
P N	2
P D	3
P Q	4
N N	5
N D	6
N Q	7
D D	8
D Q	9
Q Q	10

   So Grace can apparently make 10 combinations of two coins.

4. Check your solution. Do 10 combinations seem reasonable? They do. More than that, all the combinations are represented. Note that the table (or chart if you like) makes it easy to check all the combinations and see that there are no duplicates. Note, also, that order does not matter. That is, N Q is the same combination as Q N because the question simply asks for how many combinations, not the order of the coins.

C. Think of a Simpler Problem

You go to a party with 6 other people (besides yourself). How many handshakes does it take for all the people to shake everyone else's hand? Often a problem has a pattern that can be seen in smaller versions of the problem and can then be built up to the larger problem.

Example: Gill built a stair with wooden building blocks. They had letters of the alphabet on them. How many blocks does he need to build a stair that has 8 steps?

Use the four-step process.

1. Make sense of the problem. We need to find out how many blocks are needed to build a stair with 8 steps to it. The useless information seems to be that the blocks have alphabet letters on them. How can we solve this?
2. Make a plan. A stair with 8 steps seems pretty big. If we look at a smaller stair, we can find the number of blocks easily and thereby see the pattern that is formed.
3. Carry out the plan. If Gill makes a stair with 1 step, it will take just 1 block.

If he makes a stair of 2 steps, it will take 3 blocks.

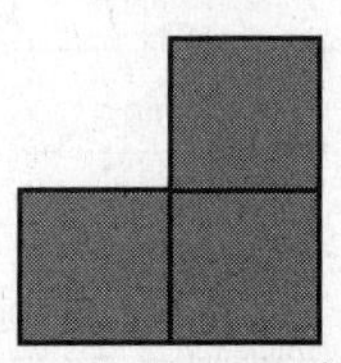

If he makes a stair of 3 steps, he will take 6 blocks.

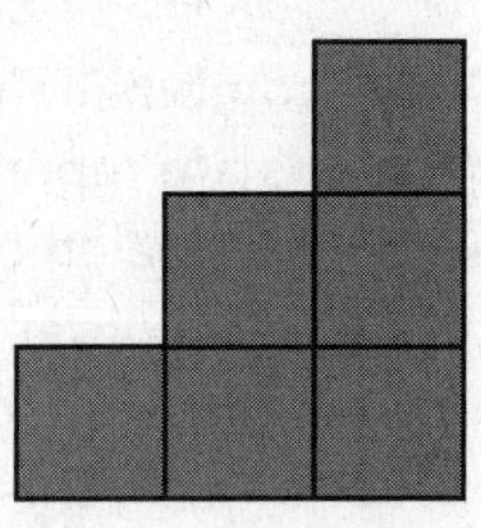

If he makes a stair of 4 steps, he will take 10 blocks.

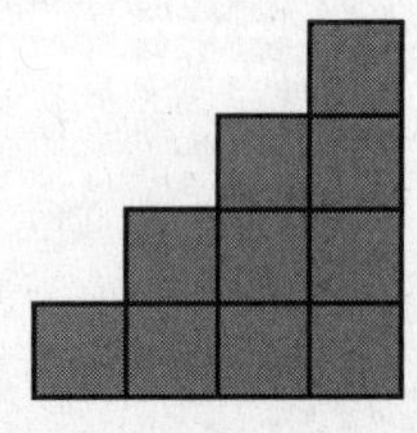

If he makes a stair of 5 steps, he will take 15 blocks.

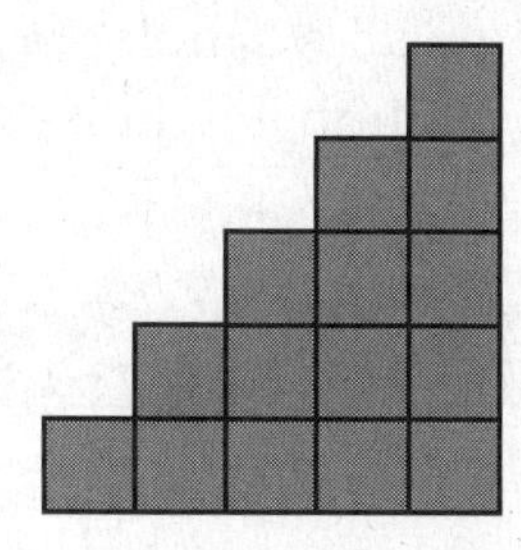

If he makes a stair of 6 steps, he will take 21 blocks.

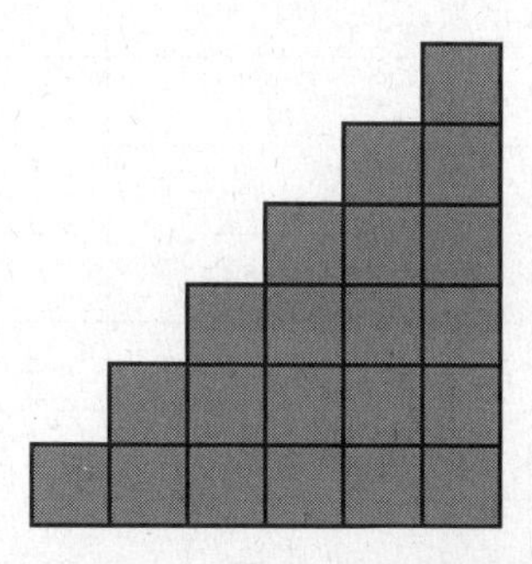

If he makes a stair of 7 steps, he will take 28 blocks.

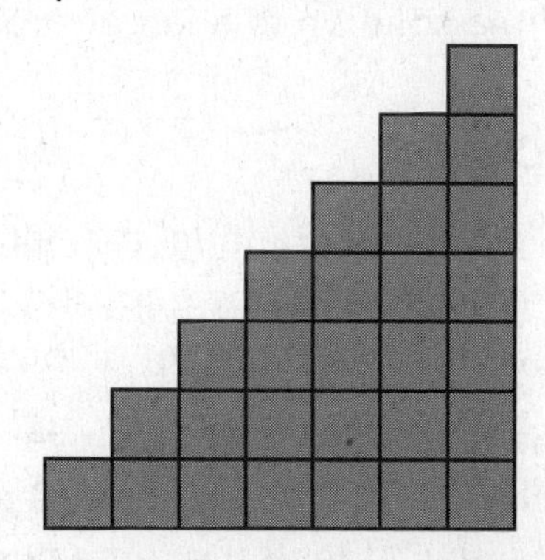

Finally, if he makes a stair of 8 steps, he will take 36 blocks.

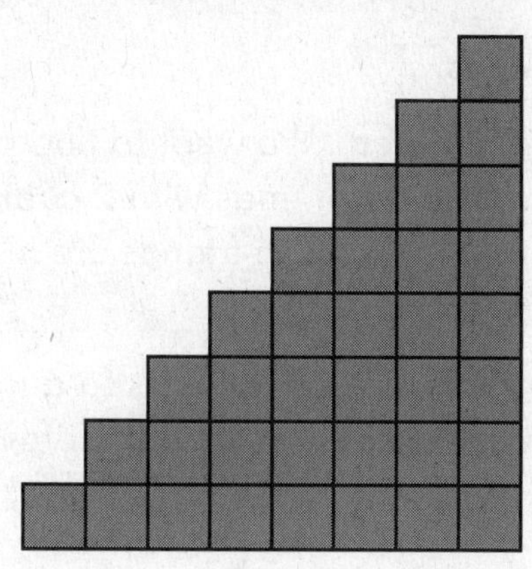

4. Check your solution. Clearly, the figures show that a stair of 8 steps will take 36 blocks. Starting with a simpler problem helps us to build up to the solution. Another way, though, might be to use a table, as with the last problem.

Steps	Blocks
1	1
2	3
3	6
4	10
5	15
6	21
7	28
8	36

Again, this table clearly shows how starting with a simpler problem allows you to find the solution to a larger problem.

D. Try Guess and Check

You want to put your collection of 28 model cars into 4 garages. Can you do it? Sometimes you can get an answer just by guessing it. The answer usually needs to be checked, though, to see if it fulfills the conditions of the problem.

Example: Elena has 7 friends from ballet class over to her house. She wants to give cookies to each of the 7 friends. Elena has 30 cookies, and she wants to keep 2 for herself. Can she offer each friend the same number of cookies with none left over?

Use the four-step process.

1. Make sense of the problem. We want to find out if there is a number of cookies Elena can give her friends with 2 cookies left over for her. The useless information is that these friends are in ballet class together. What can we do to solve this?

2. Make a plan. One way of solving this is to guess the answer and then check the solution against the conditions of the problem. That seems like a good plan.

3. Carry out the plan. We'll first remember that Elena wants 2 cookies for herself. That means that we should take away 2 cookies from the total: 30 − 2 = 28.

   We need to distribute 28 cookies out to all her friends. Let's try giving the same number of cookies to her friends that she took: 2.

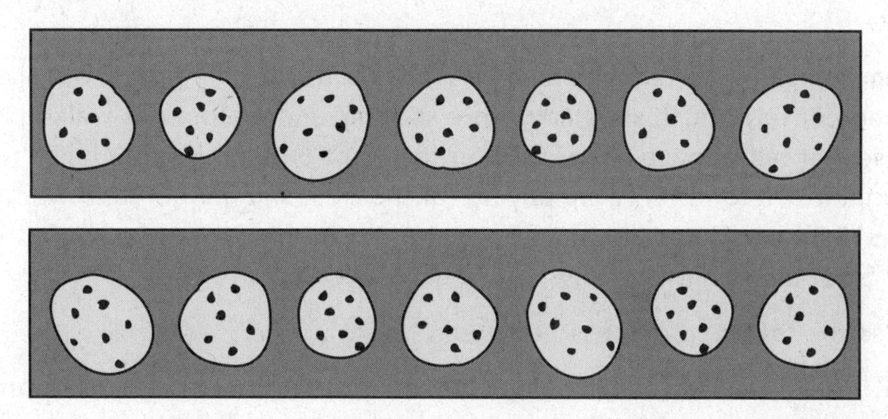

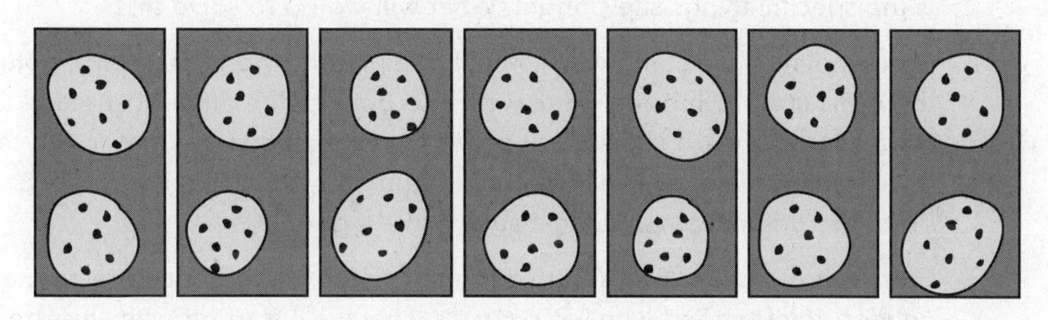

   As you can see, if we give 2 cookies to each person, we'll have 14 cookies left. However, 14 is the number we used in our first guess. So let's give another 2 cookies to each friend, like this.

   Now we have the answer. We have used up all the cookies. Each friend has the same number of cookies, 4.

4. Check your solution. We see that we have the answer. Each friend can get 4 cookies with no cookies left over. We guessed the answer and then checked it. It does seem reasonable.

E. Work Backward

You want to plan to get to school on time. School starts at 8:00 A.M. You take 10 minutes to get to school. Breakfast takes 30 minutes to make and eat, and it takes 25 minutes for you to get dressed for school. What time will you start to get to school on time? Remembering each step will help you get back to where you started.

Example: Janis started with an amount of money for shopping, but she doesn't remember how much she had before starting. Just before she walked out of the house, her father gave her $20. Then she went out and bought lunch for $7. She bought a belt for $18. While paying for the belt, she met a friend who gave her the $15 he owed her. She then bought a blouse for $23 and a hat for $15. Janis had $7 at the end of the day. How much did she start with?

Use the four-step process.

1. Make sense of the problem. We want to find out how much money she started with at the beginning of her shopping trip. The useless information is the specific things she bought. What will we do to solve this?
2. Make a plan. If we start with the final amount and work backward, doing the opposite of what Janis did, we'll end up with the amount she had at the start. For instance, if she spends money, we'll add that amount to the total. If she receives money, we'll subtract it from her total. That way, we'll back up to the original amount. That seems like a good plan.
3. Carry out the plan. We'll set up a running total, adding or subtracting as needed. Janis ended with $7, so that's what we'll start off with. Just before that, she bought a hat for $15, so

$$\begin{array}{r} \$7 \\ +\$15 \\ \hline \$22 \end{array}$$

She had $22 before she bought the hat. She bought a blouse for $23. So before that,

$$\begin{array}{r} \$22 \\ +\$23 \\ \hline \$45 \end{array}$$

She had $45 before the blouse. She received $15 from a friend before that, so

$$\begin{array}{r} \$45 \\ -\$15 \\ \hline \$30 \end{array}$$

She had \$30 before she met her friend. She bought a belt for \$18 before that, so

$$\begin{array}{r} \$30 \\ +\$18 \\ \hline \$48 \end{array}$$

She had \$48 before she bought the belt. She bought lunch before the belt for \$7, so

$$\begin{array}{r} \$48 \\ +\$7 \\ \hline \$55 \end{array}$$

She had \$55 before she got lunch. She was given \$20 before that, so

$$\begin{array}{r} \$55 \\ -\$20 \\ \hline \$35 \end{array}$$

She had \$35 before her father gave her \$20, so that's how much Janis had at the start.

4. Check your solution. To check the solution, use the starting amount we found, and do the same operations on it that the problem gives.

$$\$35 + \$20 = \$55$$

$$\$55 - \$7 = \$48$$

$$\$48 - \$18 = \$30$$

$$\$30 + \$15 = \$45$$

$$\$45 - \$23 = \$22$$

$$\$22 - \$15 = \$7$$

\$7 is what Janis ended up with, so this does check out.

F. Simulate It

Use pictures or things to act out the problem and find the solution. You have 21 students in your class, and the class is going on a field trip. Your school has vans that carry 7 students each. How can you divide up the class? Use beans and separate them into groups to help you see the solution to this one.

Example: Anastasia has a collection of 48 small, stuffed animals. She wants to put them onto her bookcase. The bookcase has lots of shelves, and each shelf can hold 6 stuffed animals. How many shelves does she need to arrange the stuffed animals if each shelf has the maximum number of animals?

Use the four-step process.

1. Make sense of the problem. We need to figure out how many shelves we will need to arrange the animals. We will be arranging them in groups of 6. How can we solve this?
2. Make a plan. If we picture the stuffed animals in groups of 6, we can find out how many shelves we need. So drawing a picture or using beans or some other manipulative seems like a good way to solve this.
3. Carry out the plan. We draw the picture that arranges the stuffed animals in groups of 6 until all the stuffed animals are used up.

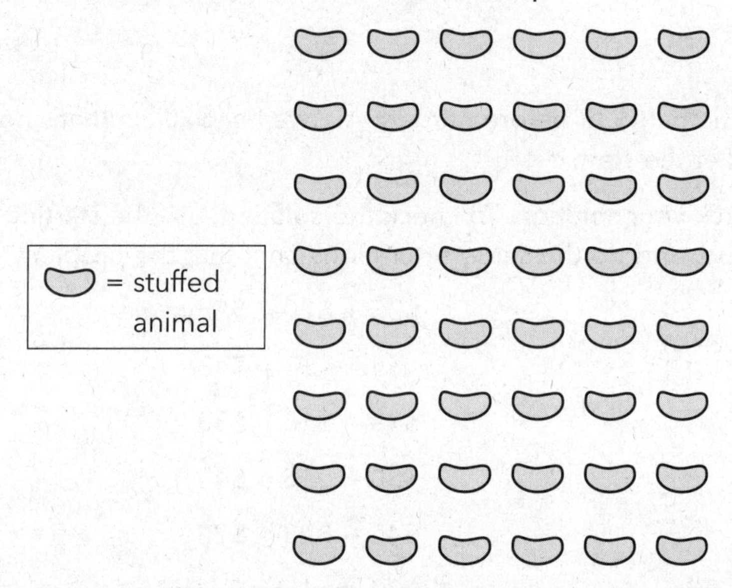

As you can see, we can arrange the stuffed animals in 8 shelves of 6 each. This arrangement will store all the animals, with no partially full shelves. Using mathematical language, this is: $8 \times 6 = 48$.

4. Check your solution. Did we store all 48 stuffed animals? Yes, and there were no shelves that were partially filled with stuffed animals. We used 8 shelves to store the stuffed animals with 6 on each shelf. Another way to solve this would be to start with 48 beans and take 6 away at a time, keeping track of how many times we take 6 away.

Another Example: 57 boys and 72 girls entered a spelling bee. How many students entered the spelling bee in all? How many more girls than boys entered the spelling bee?

1. Make sense of the problem. We want to find two values: how many more girls than boys entered the spelling bee and the total number of students who entered the spelling bee. How can we solve these?
2. Make a plan. Drawing a picture and comparing the two numbers (the boys and the girls) could help us find the answers.
3. Execute the plan. We draw two bars that represent the two groups of students.

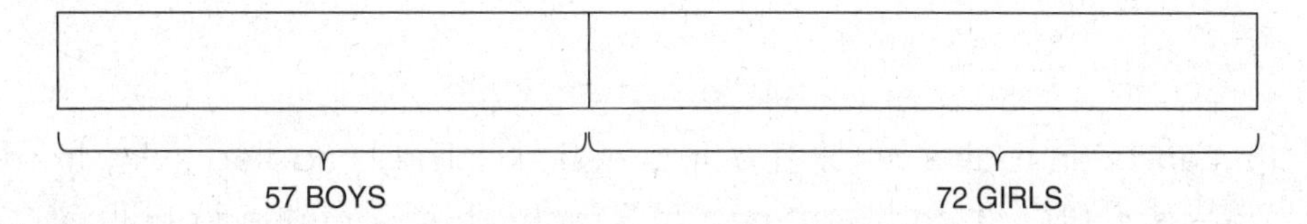

We see that 57 boys and 72 girls give us: 57 + 72 = 129. The total number of students entering the spelling bee is 129.

Next we find how many more girls than boys entered the spelling bee. We compare the two bars by setting one on top and one on the bottom.

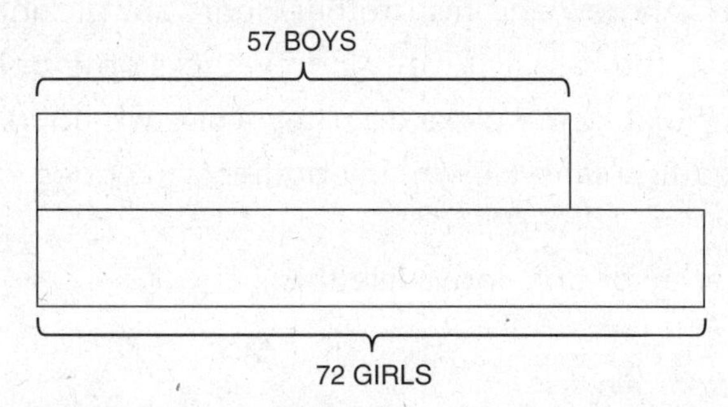

We see, then, that 72 − 57 = 15. There were 15 more girls than boys who entered the spelling bee.

4. Check your solution. Do the answers make sense? As for the total number of students, 57 + 72 = 129, so this checks. Since there were 72 girls and 57 boys, 72 − 57 = 15, meaning that 15 more girls than boys entered the spelling bee.

Problem Solving Exercises

1. Greg is having several friends over for a barbeque in the afternoon. He will be cooking hamburgers, hot dogs, ribs, and chicken drumsticks on the grill. Hamburgers take 20 minutes to cook. Ribs take 35 minutes to cook. Hot dogs take 10 minutes to cook. Chicken drumsticks take 30 minutes to cook. He wants to take all the items off the grill at 3:00 P.M. If he has a grill that will fit all the different meats he is cooking, when should he put each meat onto the grill to ensure that they will all come off at 3:00 P.M.?

2. Some members of the Gaskin family (including some aunts and/or uncles) went to River City Zoo recently. The children invited some of their friends, so that there were more children than adults. Altogether, the group paid $104. If the admission price for children is $7 and the admission price for adults is $12, how many children and how many adults went on the Gaskin's trip to the Zoo?

3. Phil and Ben took a kayak trip in Long Lake. They paddled away from Hale Dock at 8:00 A.M. and went east at 5 mph. They continued until noon, at which time they pulled onto shore for lunch. Then they got back into their kayaks at 1:00 P.M. and paddled west at a speed of 4 mph, paddling until 5:00 P.M. How close were they to the dock by then?

4. For Earth Day, Andrew and his brother Henry are picking up plastic bottles and cans along the Passaic River. Andrew picks up 6 bottles/cans for every 5 bottles/cans that Henry picks up. They work two hours, and together they collect 132 bottles/cans. How many bottles/cans does each boy pick up?

5. A triangular array of dots looks like this:

How many dots would there be in a triangular array that has 10 dots on each side?

Answers to Problem Solving Exercises

1. Greg is planning his barbeque. He needs to decide when to put the various items onto his grill. It would seem that if we worked backward from 3:00 P.M., we'll get our answer.

Hamburgers:	Ribs:	Hot Dogs:	Chicken:
3:00 P.M.	3:00 P.M.	3:00 P.M.	3:00 P.M.
− 20 min.	− 35 min.	− 10 min.	− 30 min.
2:40 P.M.	2:25 P.M.	2:50 P.M.	2:30 P.M.

If Grey starts at 2:25 P.M. by putting the ribs on, and then puts the chicken on at 2:30 P.M., the hamburgers on at 2:40 P.M., and finally the hot dogs on at 2:50 P.M., he'll be able to take all of them off the grill at 3:00 P.M.

2. Admission is $7 per child and $12 per adult. So some combination of 7s and 12s that add up to 104 should give us our answer. One restriction we need to remember is that there are more children than adults. Guess and check appears to be a good way to solve this.

   Let's try 7 children and 6 adults:

$$\begin{array}{r} 7\times7=49 \\ 12\times6=\underline{72} \\ \$121 \end{array}$$

   That's too much. We see that we'll get an odd number if we multiply 7 by an odd number. We need to multiply 7 by an even number to get 104.

   Let's try 8 children and 5 adults:

$$\begin{array}{r} 7\times8=56 \\ 12\times5=\underline{60} \\ \$116 \end{array}$$

   Again, that's too much. It's too much by just $12, though. Let's drop one of the adults.

$$\begin{array}{r} 7\times8=56 \\ 12\times4=\underline{48} \\ \$104 \end{array}$$

   We've got it. Do you see how, in each case, the previous guess helped to narrow choices for the next guess and helped us find the answer? Always try to think of how the last guess helps you move toward the next guess. This is one of the most powerful aspects of the guess and check method: its ability to narrow our choices, getting us closer to the answer with each guess.

3. Phil and Ben are kayaking in Long Lake. We need to find out how far they have kayaked and how far they are from their starting point. Drawing a picture would seem to work for this one but not before we calculate how far they have paddled one direction and then the other.

   They paddled eastbound from 8:00 A.M. until noon. That is 4 hours. They went 5 mph during that time: $4 \times 5 = 20$ mi. They traveled 20 miles east. Then they traveled from 1:00 P.M. until 5:00 P.M. That is 4 hours. They traveled at a speed of 4 mph during that time: $4 \times 4 = 16$ mi. They traveled 16 miles west. So let's draw a picture of this:

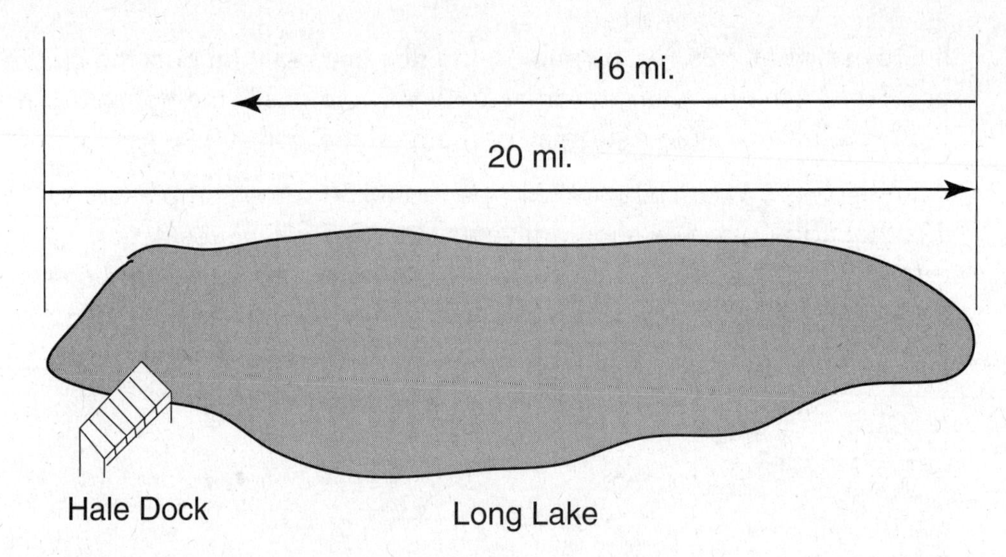

As you can see, Phil and Ben are not at their original starting place. To find out how short they are from it, subtract 16 from 20: 20 − 16 = 4. They are 4 miles short of their starting point. This is a good example of a problem that can be solved by combining a picture (or drawing) and a few simple equations.

4. Andrew picks up 6 bottles/cans for every 5 bottles/cans that Henry picks up. We know that a total of 132 bottles/cans are picked up. We want to know how many each boy picked up. If we make up a table, we can keep track of how much each boy picks up and of the running total.

Andrew	Henry	Total
6	5	11
12	10	22
18	15	33
24	20	44
30	25	55
36	30	66
42	35	77
48	40	88
54	45	99
60	50	110
66	55	121
72	60	132

So Andrew picks up 72 bottles/cans, and Henry picks up 60 bottles/cans. This gives us a total of 132 bottles/cans.

5. We need to find out how many dots are in a triangular array that has 10 dots on a side. A chart, along with a picture, would help us to see the solution.

As you can see, a triangle with 1 dot on a side has 1 dot total. An array with 2 dots on a side has 3 dots total. An array with 3 dots on a side has 6 dots total. Let's look at this in a table.

Number of Dots Per Side	Total Number of Dots
1	1
2	3
3	6
4	10
5	15
6	21
7	28
8	36
9	45
10	55

So a triangular array of dots that has 10 dots on a side contains 55 dots.

Test Your Skills

1. In a sequence of steps, we know all of the steps except the first one. What is the first number? (Note: The order of operations rules are not being followed here.)

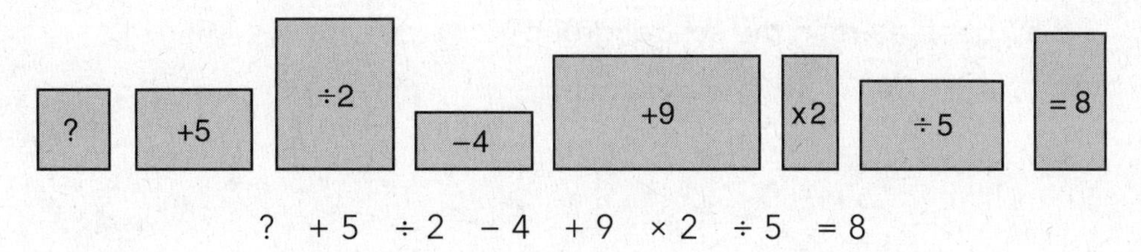

? + 5 ÷ 2 − 4 + 9 × 2 ÷ 5 = 8

2. Zack wants to put a fence around his mom's vegetable garden. The garden is in one corner of the yard and is shaped like a trapezoid.

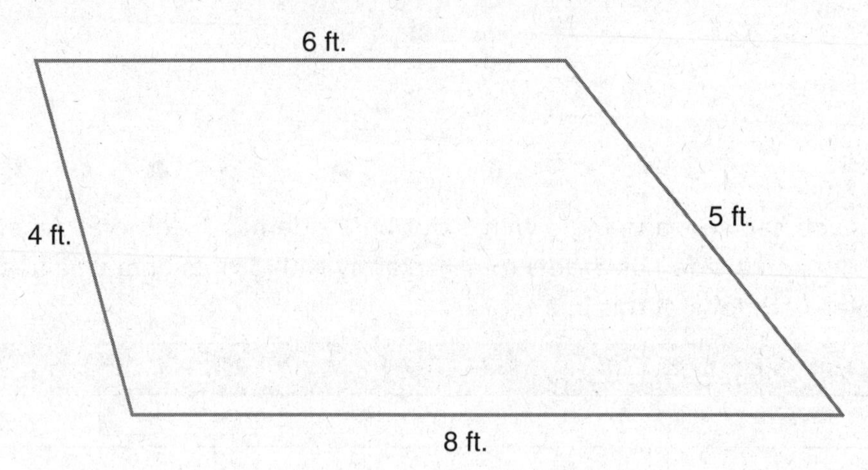

How many feet of fence will Zack need to build the fence?

3. The lockers on the first floor in Third River High School are numbered 100–200. How many lockers have a 6 in them?

4. Consider this number sequence: 23, 28, 26, 31, 29, 34,...What is the next number in this sequence?

5. Peter wants to plant a tree in his front yard. The yard is 50 feet wide. The dirt ball around the roots of the tree is 4 feet in diameter. How far from each side of the yard must he plant the tree to center it in the front yard?

6. Mickey has a quarter, a nickel, and two pennies in his pocket. How many different amounts of money can he make using the available coins? (For instance, one nickel and one penny would make 6 cents.)

7. The battleship New Jersey gives a break to school groups. For every 8 tickets sold, the group gets 1 free ticket. How many free tickets will a school receive if 50 people are in the school group?

8. Consider the following graph:

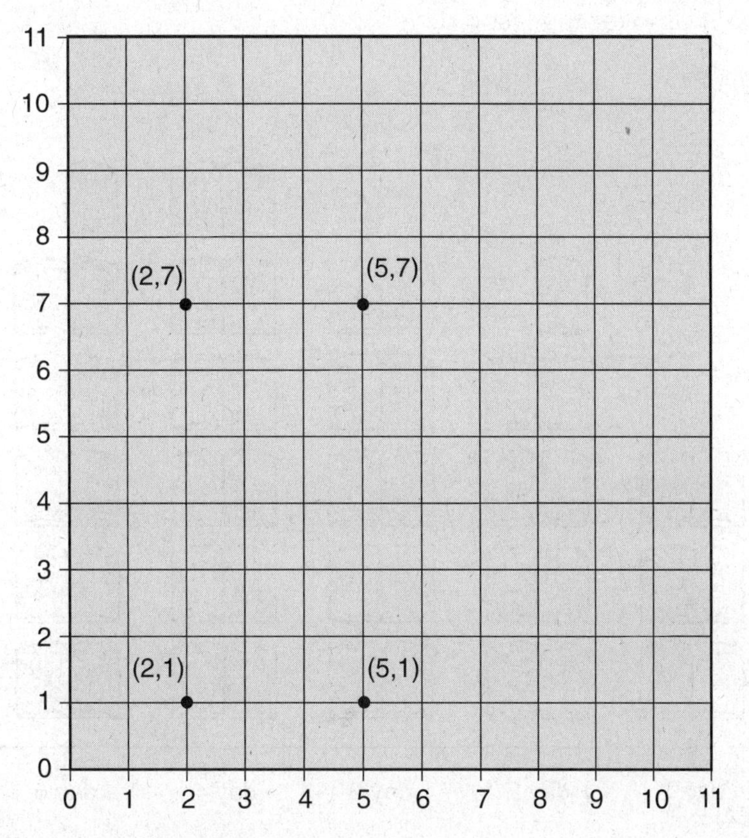

A. If these four points were connected, what 2-dimensional geometric shape would be formed?

B. What is the perimeter of the shape?

C. If you were to slide this shape 4 units up, where would the points be plotted on the graph?

9. The following pictograph shows the amount of books sold at Bob and Ray's Book Shop over the last 6 days.

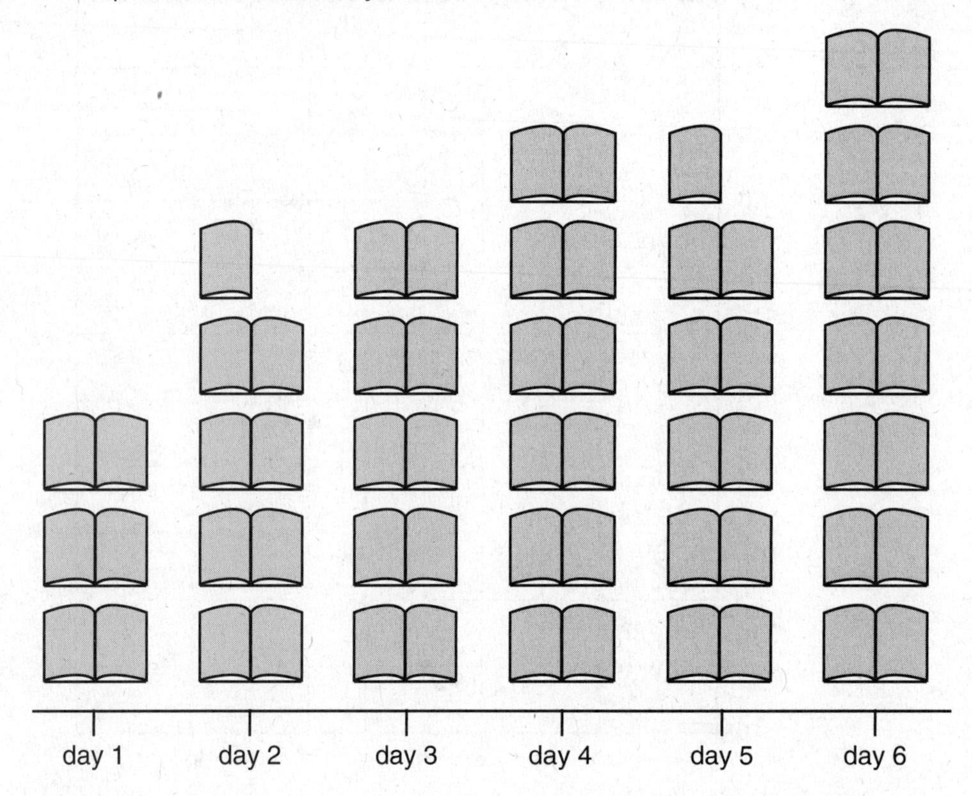

Each book represents 10 books.

A. How many books did Bob and Ray's Book Shop sell on the fourth day?

B. On what day did the sales go down?

C. How many books did Bob and Ray's Book Shop sell during the entire 6 days?

10. Martina is getting outfits together for her Caribbean vacation. She has 4 blouses (1 red, 1 gray, 1 blue, and 1 white) and 3 pairs of pants (1 navy, 1 dark gray, and 1 tan). How many different outfit combinations can she make for herself if she wears 1 blouse and 1 pair of pants? Make up a tree or other diagram to show your solution.

If she had 2 pairs of shoes (1 pair of sneakers and 1 pair of penny loafers), how many combinations of 1 blouse, 1 pair of pants, and 1 pair of shoes can she make? Use a tree or other diagram to show your solution.

Answers to Problem Solving and Other Math Skills

For all of these problems we'll use the four-step process.

1. **a) Make sense of the problem:** We're trying to find the beginning number in a sequence of operations. What are the facts? We have the sequence of numbers, and the operations, but not the beginning number. We know all the operations and all the numbers, except the beginning one. How can we do this?

   b) Make a plan: Can we draw a picture? That won't help. Can we look for a pattern? Not really, because there is no pattern to see. Can we work backward? Well, that might help, but we need to know what we started with. Can we guess and check? Yes, that seems to be the right way to proceed.

   c) Carry out the plan: So, let's try a starting number. The number should be odd, because we add 5 right away, and then divide by 2, so we need an even number. Starting with an odd number will insure an even number when we have to divide by 2. Let's try 9:

   $9 + 5 = 14 \div 2 = 7 - 4 = 3 + 9 = 12 \cdot 2 = 24 \div 5 =$ Oh, oh. We can't divide 24 by 5 evenly, and our final answer is 8. We know that 9 does not work, but we need a number larger than that number. Let's try 15: $15 + 5 = 20 \div 2 = 10 - 4 = 6 + 9 = 15 \cdot 2 = 30 \div 5 = 6$. We're not far away from 8 (just two off). But, by using a multiple of 5, we seemed to get a whole number for an answer. Let's try 25: $25 + 5 = 30 \div 2 = 15 - 4 = 11 + 9 = 20 \cdot 2 = 40 \div 5 = 8$! We've found it. Did you see that we gleaned clues as we went along for how to "choose smarter" on the next pick? Try to do that all the time.

   d) Check your answer: does this work? Yes, because when you plug the number in you'll get 8. Does this make sense? Yes, it does. Are there other numbers that would work? Probably not. So, this works.

2. **a) Make sense of the problem:** We're trying to find the perimeter to get the proper amount of fence. The garden is shaped like a trapezoid. How will we do this?

   b) Make a plan: This is not a difficult problem. If we get the perimeter, we'll know the amount of fence we need. We'll add up all the legs of the trapezoid. It might be helpful to draw a diagram of the garden.

   c) Carry out the plan: We have $4 + 6 + 5 + 8 = 23$. That's 23 feet. When you know units, you should put them in.

   d) Check your answer: Does this make sense? Yes it's reasonable that 23 feet of fence will adequately fence in this garden. Are there other answers that will work? Not really, because we need an exact amount, here. So this works.

3. **a) Make sense of the problem:** We need to know how many numbers between 100 and 200 have a 6 in them. What are the facts? The numbers from 100–200. How will we do this?

   b) Make a plan: Can we guess? That won't work, because we'll have no good way to check it, except to look at all the numbers, which will take a long time.... Can we make a picture? A picture won't solve the problem. Can we look for a pattern? Yes, that probably will get the answer for us. So let's try a pattern.

   c) Carry out the plan: How will we set up the plan? Let's look at the first ten numbers. 100, 101, 102, 103, 104, 105, 106, 107, 108, 109, 110. So, for the first ten numbers, a 6 appears once. So, for each group of ten numbers, a 6 will appear once. Is this true? Let's check the next ten numbers: 111, 112, 113, 114, 115, 116, 117, 118, 119, 120. Yes, that works. There are ten groups of these tens, so in the ten groups of tens from 100–200, there are ten numbers that have a 6 in them. Is that right? Oh, wait! When we get to the numbers 160 to 169, all ten numbers have a 6 in them. OK, so there are ten more numbers: 10 + 10 = 20. Is that right? Well, not really, because one of those numbers, 166, is being counted twice, so we need to take one away from that 20. There are 19 numbers between 100 and 200 that have a 6 in them.

   d) Check your answer: Does this make sense? Well, we looked at the numbers, found a pattern, and followed that pattern. It's reasonable that there are 19 numbers that have a 6 in them.

4. **a) Make sense of the problem:** What is the required answer? We want the last number in the sequence. What are the facts? We have the sequence of numbers. How can we get this? Figure out the pattern (how we get from one number to the next) and then apply that rule to the last number to get the required number.

   b) Make a plan: How will we solve this? Searching for a pattern seems to be the best way to approach this problem. We'll examine each number, and see how to get from one number to the next, and see what the pattern is to do that. Then we'll try to apply that pattern to the next number in the sequence.

   c) Carry out the plan: Let's look at all the numbers in the sequence:

   23 28 26 31 29 34...

This is a very strange sequence. It goes up, then down, then up again. What number does it go up by? 5, initially, then down 2, then up 5, then down 2, then up 5 again. So, the next number in the sequence would be down by 2 or 32.

d) Check your answer: Does this sound reasonable? Yes, it does. The pattern is: 5 up, 2 back, and repeat. So following that pattern, the next number will be the last number minus 2, or 32.

5. **a) Make sense of the problem:** What is the required answer? We want to find out the distance from each side of the front yard to plant a tree. What are the facts? We have the diameter of the dirt ball, and the width of the yard. How can we get this? We can draw a picture to find out where to plant the dirt ball of the tree.

   b) Make a plan: We'll draw a picture so that we can figure how far from each side Peter needs to put the dirt ball so that it is in the center of the yard. Peter might also look at how to center it using only numbers.

   c) Carry out the plan:

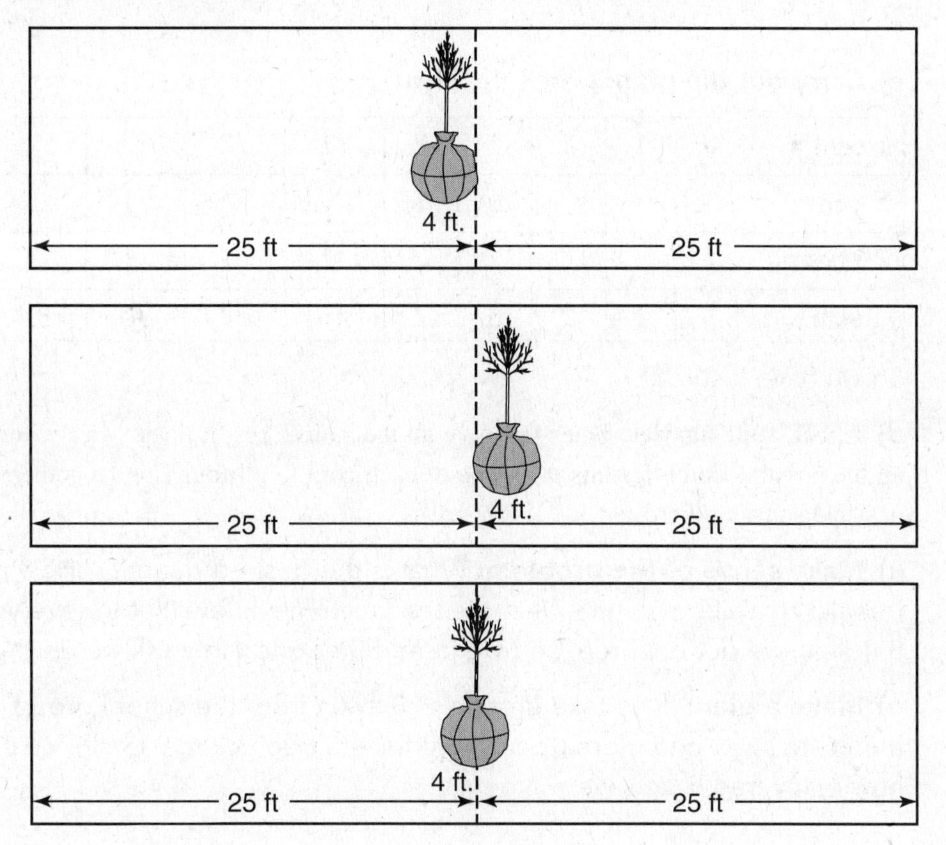

You see that if Peter puts the dirt ball at 25 ft., it would extend 4 ft. on one side or the other, so it wouldn't be centered. He must place it so that 2 ft. are extended on each side of the center, so that it is centered.

There is another way. The yard is 50 ft. long, and half of that is 25. The dirt ball is 4 ft., so cut it in half, giving 2 ft. on each side. So the dirt ball will start at 23 ft. and end at 27 ft. on the number line, and that will center it.

d) Check your answer: Does this answer the question? Yes, it does. It places the dirt ball exactly in the center of the yard. Peter saw two ways of doing it here, and many problems can be done in more than one way.

6. **a) Make sense of the problem:** What is the question being asked? How many different sums of money can be made with four coins? What are the facts? Mickey has 2 pennies, a nickel, and a quarter. Note that using one or the other penny will not give another sum of money. Mickey wants only to find different sums.

b) Make a plan: Will a picture help? Not really. A chart probably will be best, because it will show all the possibilities very clearly. So Mickey will make a chart.

c) Carry out the plan: Here's the chart

1 cent	1	2	1	2	1	2		1	2
5 cents			1	1			1	1	1
25 cents					1	1	1	1	1
sum	1	2	6	7	26	27	30	31	32

So we have 9 sums.

d) Check your answer: Does this give all the sums? Yes, it does. Mickey looked at all the possible combinations from one of each coin to all four coins together. All the possible sums are here.

7. **a) Make sense of the problem:** What is the desired quantity? How many free tickets will be given? What are the facts? We know that for every 8 tickets, we get one free, and we need 50 tickets. How will we do this?

b) Make a plan: If we take groups of 8 away from the school group, and then take away one more (to account for the free ticket, 9 in all), we'll find out how many free tickets we will get.

c) Carry out the plan: We start with 50:

$$50 - 9 = 41$$

$$41 - 9 = 32$$

$$32 - 9 = 23$$

$$23 - 9 = 14$$

$$14 - 9 = 5$$

5 – 9 ? We cannot take 9 from 5, so we're done. We don't have another 8 students to get a free ticket, so we'll get 5 free tickets.

d) Check your answer: Does this answer the question? Yes, it does. See how repeated subtraction will get us to our answer.

8. **a) Make sense of the problem:** What are we to find? We need to name the shape formed by the four points, and we need to find the perimeter of the shape. Then, we need to slide the figure up four units, and calculate where the points would be that form its corners.

b) Make a plan: The best way to solve this is to plot the points on a graph and connect them. Then, we can see what shape is made.

c) Carry out the plan:

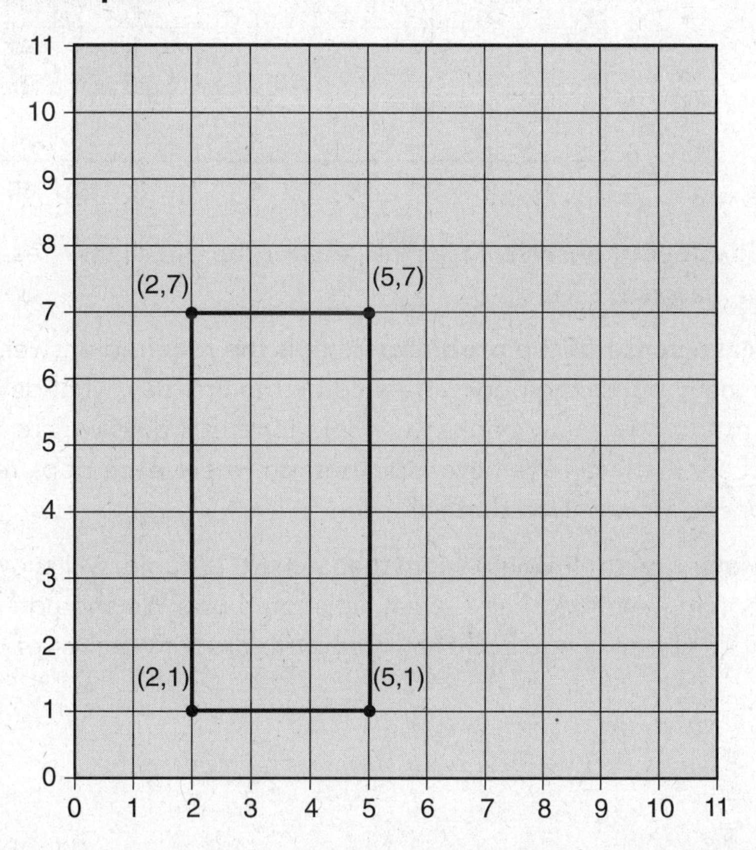

A. We see that the figure made is a rectangle.

B. We see that the rectangle's width is 3 units, and the length is 6 units, so the perimeter is 3 + 6 + 3 + 6 = 18.

C. If we slide the rectangle up four units, we get this

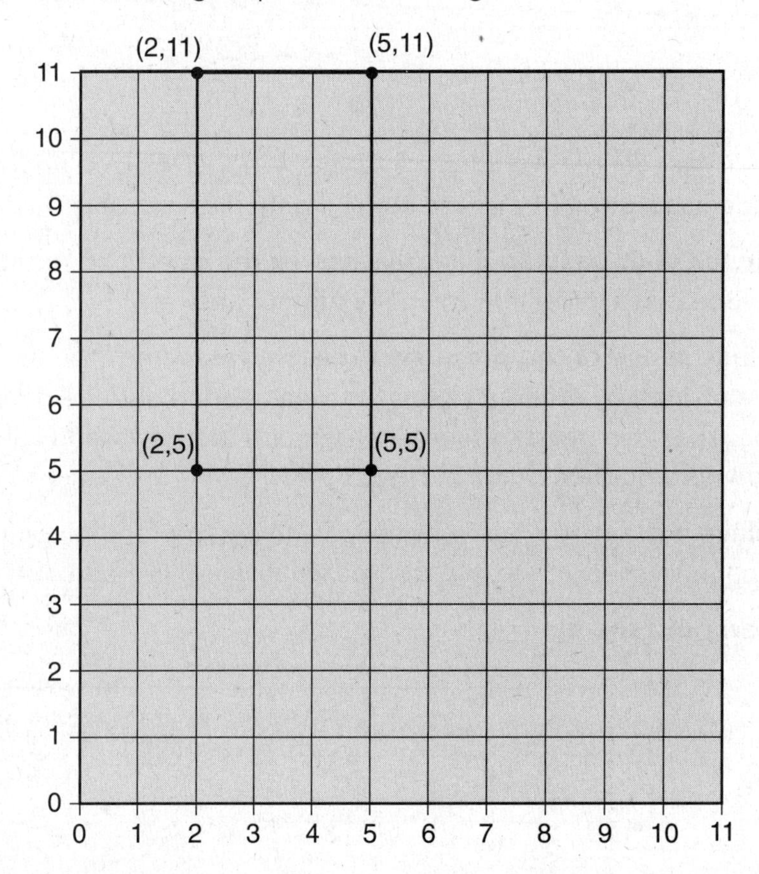

d) Check your answer: Does this answer the questions? Yes, it does. Does it seem reasonable? Yes.

9. **a) Make sense of the problem:** What is the required answer? It's to know how many books the bookstore sold on the first day, what day the sales went down, and how many books were sold for all six days we are looking at. What are the facts? We have a pictograph where each book represents 10 books. How will we do this?

   b) Make a plan: It would seem that a chart, or table, will show the amounts sold more clearly. A chart is what we should use. We should expand the number of books out, however, giving the proper number of books for each.

c) Carry out the plan: Here is the chart:

Day	Books Sold
1	30
2	45
3	50
4	60
5	55
6	70

So, the answers to the three questions are:

A. Bob and Ray sold 60 books on the fourth day.

B. Sales went down on the fifth day.

C. The total amount of books sold in the six days was:

$$30 + 45 + 50 + 60 + 55 + 70 = 310 \text{ books}$$

d) Check your answer: Does this answer all the questions? Yes, it does. It's much easier to read when you put the information in a table.

10. **a) Make sense of the problem:** What is the required answer? How many different combinations of outfits can Martina make? What are the facts? The number of blouses Martina has (4), and the number of pants she has (3). How will we do this?

b) Make a plan: The way we can solve this is to make a tree diagram.

c) Carry out the plan: Martina will make a diagram that matches up all the blouses with all the pants.

You see that there are 12 combinations of blouses with pants.

If she had two pairs of shoes, how many combinations of blouses, pants, and shoes will Martina have?

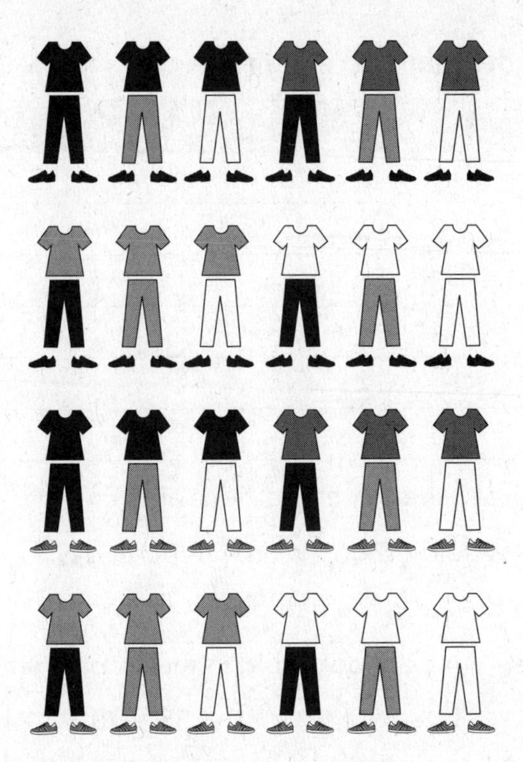

You see that there are 24 combinations of blouses, pants, and shoes.

d) **Check your answer:** Does this make sense? Yes, it does. You can see clearly, with a tree diagram, *all* the combinations that are available to Martina when she goes on vacation.

Operations and Algebraic Thinking

CHAPTER 3

MULTIPLICATION

DEFINITIONS/VOCABULARY

Associative Property of Multiplication: This property states that when the grouping of factors is changed, the product remains the same. (For example, 3 × 5 × 2 can be found by multiplying 3 × 5 = 15 and then 15 × 2 = 30, or by multiplying 5 × 2 = 10 and then 3 × 10 = 30.)

Commutative Property of Multiplication: This property states that you can multiply two factors in either order and get the same product. (For example, if 6 × 4 = 24 is known, then 4 × 6 = 24 is also known.)

Distributive Property of Multiplication: This property states that multiplying a sum by a number is the same as multiplying each addend by the number and then adding the products. (For example, since 8 × 5 = 40 and 8 × 2 = 16, one can find 8 × 7 as 8 × (5 + 2) = (8 × 5) + (8 × 2) = 40 + 16 = 56.)

Factor: A number that is multiplied by another number to get a product.

Product: The result when you multiply together at least 2 factors.

Multiplication is a process very similar to addition. In fact, it is addition on an extended scale using equal groups. For example, 4 + 4 + 4 + 4 + 4 is the same as 4 × 5 because there are five 4's being added together. Also like addition, multiplication has a set of basic facts that must be memorized. This section of the book focuses on strategies for memorizing multiplication facts and strategies for determining the answers to basic multiplication facts. At this grade level, these basic facts include the factors through 10, including 10 × 10. Ultimately, memorizing these multiplication facts is very important because a good portion of third grade, and every grade after that, will rely on these as a foundation for working through more complex and multistep concepts.

Multiplication Facts

A few strategies can be used either to memorize or to assist in computing the basic multiplication facts. Let's start by looking at a few strategies that can be helpful with all forms of multiplication involving two single-digit factors.

For all multiplication facts, you can count or draw equal groups and then skip count to find the total. For example, if I asked you what 5 × 3 is, you could skip count by 5s three times: 5, 10, 15. So 15 is the answer. You could also skip count backward if that is easier for you. Count by 3s five times: 3, 6, 9, 12, 15. Again we see that 15 is the answer. This is known as the commutative property of multiplication. It states that you can multiply two factors in any order and get the same product.

Another strategy is known as the distributive property of multiplication. This strategy states that multiplying a sum by a number is the same as multiplying each addend by the number and then adding the products. For example, let's say you are faced with the problem of multiplying 8 × 7. If you do not know the answer to this fact, you can break it down into two separate facts and then add the products. We could break it down to 8 × 3 and 8 × 4 (breaking the 7 into a 3 and a 4 while leaving the 8 alone): 8 × 3 = 24, and 8 × 4 = 32. So we take those two products and add them together: 24 + 32 = 56. So 56 is the product of 8 × 7.

Additionally, if these methods of mental math are too difficult you can always draw out the problem. One way to do that is using equal groups. In the question 3 × 6 (or 3 groups of 6), you might solve by drawing equal groups and then skip count to find the total.

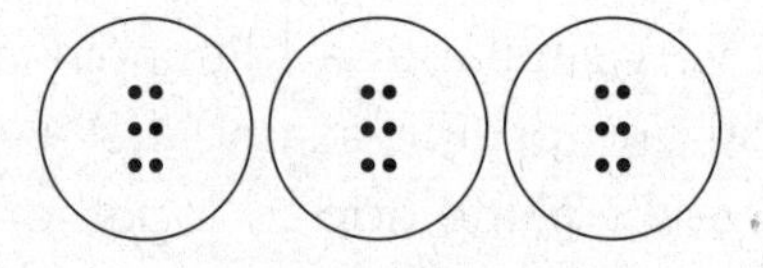

If we skip count, we end up with 6, 12, 18. This tells us that 3 × 6 (or 3 groups of 6) is equal to 18.

Similar to skip counting is multiplication using arrays. For example, if you were multiplying 4 × 7, you would make 4 groups of 7 (using dots, counters, or any other appropriate manipulative) and then add them to get a total of 28. An example would be an array for 4 × 7 that shows 4 columns of 7 dots in each or 7 columns of 4 dots in each (per the commutative property of multiplication).

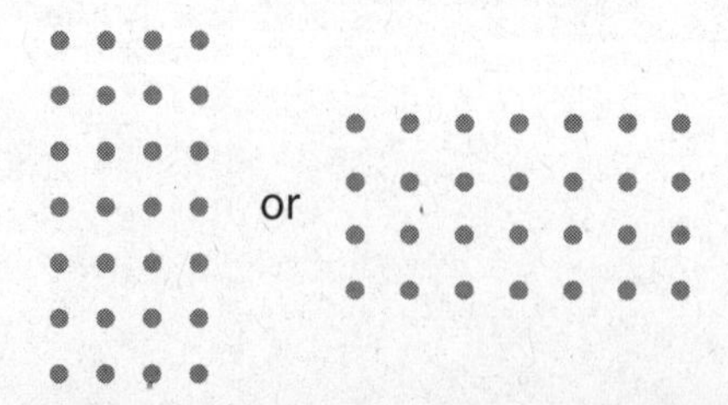

Strategies for Memorizing Specific Multiplication Facts	
× 0	Any number times 0 is 0 (for example, 8 × 0 = 0).
× 1	Any number times 1 is itself (for example, 9 × 1 = 9).
× 2	Any number times 2 is double itself (for example, 6 × 2 = 12, the same as 6 + 6).
× 3	In any multiple of 3, the product's digits will add up to a multiple of 3 (for example, 3 × 6 = 18; if we add 1 + 8, we get 9, which is a multiple of 3).
× 5	Any number times 5 will end in either 5 or 0 (for example, 6 × 5 = 30 and 5 × 7 = 35).
× 6	If you multiply 6 by an even number, that factor and the product will both end in the same digit (for example, $6 \times \underline{2} = 1\underline{2}$, $6 \times \underline{4} = 2\underline{4}$, $6 \times \underline{6} = 3\underline{6}$).
× 9	There is a neat trick for multiplication times 9s up to 9 × 10. Let's say we're multiplying 9 × 8. Hold both of your hands in the air and spread apart all 10 of your fingers. (When we refer to fingers we are including thumbs.) This trick works ONLY for the 9s facts. Since we're doing 9 × 8, start at your left pinky and count in your head to your 8th finger. You should end at your right index finger (next to your thumb). Put that finger down and keep all the others up. Now look how many fingers are to the left of the finger you just put down. You should see 7. Now look how many fingers are to the right of the finger you just put down. You should see 2. Put those two numbers (7 and 2) together and you have your answer to 9 × 8, which is 72.
Other	There is a trick you can do when multiplying by rounding up the number and then subtracting backward. For example, if you did not know that 9 × 7 = 63, you could do 10 × 7, which is a lot easier, and get 70. Then subtract a 7 and get 63. This works because originally we asked how much nine 7s were. If instead you find ten 7s, you can just subtract a 7 afterward and still get the correct answer. Here's another example. 5 × 9 = 45. If you didn't know this immediately and couldn't figure it out by skip counting, you could instead multiply 5 × 10 and get 50. Then just subtract a 5 (because you found ten 5s instead of nine 5s) and get 45.

Multiplication Facts Exercises

Use one of the strategies you have learned to solve the following problems:

1. What is 5 × 7?

- O A. 35
- O B. 40
- O C. 30
- O D. 23

2. What is 9 × 2?

- O A. 16
- O B. 18
- O C. 13
- O D. 9

3. Which is 9 × 7?

- O A. 63
- O B. 61
- O C. 72
- O D. 54

4. What is 7 × 7?

- O A. 42
- O B. 49
- O C. 7
- O D. 1

5. What is the missing factor in the equation $45 = n \times 9$?

- O A. 5
- O B. 13
- O C. 36
- O D. 6

6. What is 8 × 0?

- O A. 8
- O B. 1
- O C. 0
- O D. 16

7. The addition sentence 7 + 7 + 7 + 7 + 7 has the same value as:

- ○ A. 7 × 7
- ○ B. 7 − 5
- ○ C. 7 × 5
- ○ D. 14

8. A box of pencils costs $2.00. What is the total cost if 3 boxes are purchased?

- ○ A. $7.00
- ○ B. $5.00
- ○ C. $6.00
- ○ D. $1.00

9. A school has 9 classrooms. Each classroom has 6 windows. How many windows does the school have altogether?

- ○ A. 54
- ○ B. 45
- ○ C. 15
- ○ D. 63

10. Victoria wants to purchase 4 new dolls for her dollhouse. Each doll costs $8. How much will the dolls cost altogether?

- ○ A. $12.00
- ○ B. $4.00
- ○ C. $24.00
- ○ D. $32.00

11. Angelina purchased 4 packs of muffins for her class's holiday party. If each pack contained 6 muffins, how many muffins did she purchase altogether?

- ○ A. 12
- ○ B. 24
- ○ C. 10
- ○ D. 2

12. In the schoolyard, 3 trees need trimming. The tree trimmer arrives and removes 9 branches from each tree. How many branches did the tree trimmer remove altogether?

- ○ A. 12
- ○ B. 6
- ○ C. 27
- ○ D. 36

13. Which array represents the expression 4×7?

○ A.

○ B.

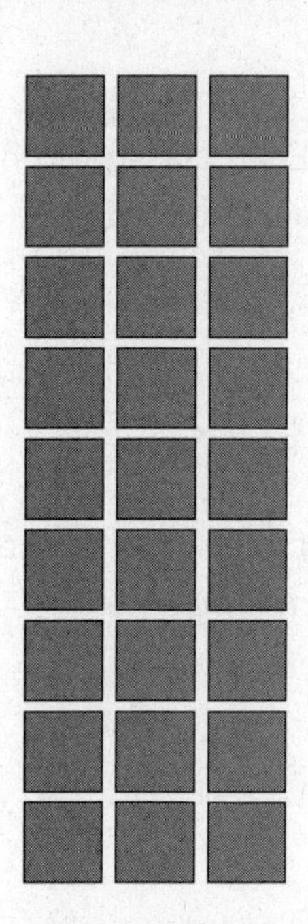

○ C.

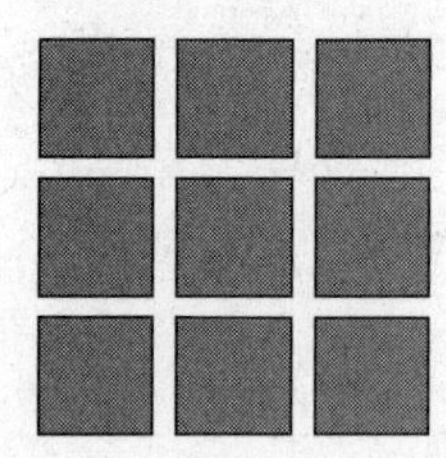

○ D.

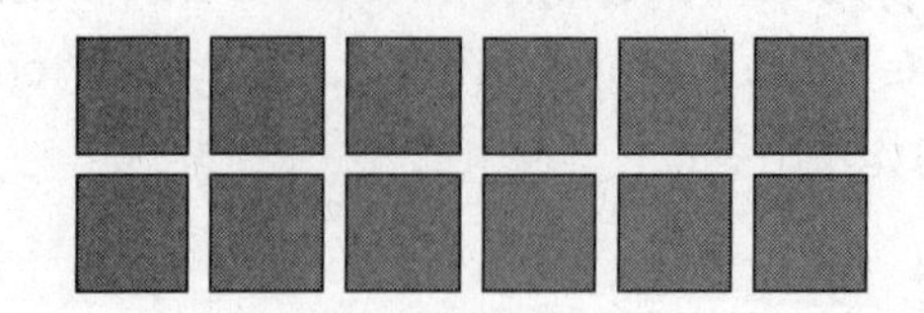

14. Write a multiplication expression that matches the array shown.

Explain your answer: --

Answers to Multiplication Facts Exercises

1. **A.** If you count by 5 seven times, you will see the answer is 35.
2. **B.** If you count by 2 nine times, you will see the answer is 18.
3. **A.** If you do the 9s trick using your fingers, you will see the answer is 63.
4. **B.** $7 \times 7 = 49$. You could memorize this fact or create an array to solve it.
5. **A.** In the equation $45 = n \times 9$, the missing factor is 5. If we replace the n with a 5, we can clearly see that 45 is equal to 5×9. Remember that the equals sign means the value of the expression on the left must be the same as the expression on the right.
6. **C.** Any number times zero is zero.
7. **C.** The addition sentence shows 5 sevens being added together. This is the same as 7×5.
8. **C.** One box of pencils costs $2.00. If you purchase 3 boxes, you solve by multiplying 2 × 3 = $6.00.
9. **A.** If the school has 9 classrooms and each classroom has 6 windows, multiply 9×6 to get to an answer of 54 windows.
10. **D.** If each doll costs $8.00 and Victoria wants to purchase 4 dolls, solve using multiplication: $8.00 × 4 = $32.00.
11. **B.** Each package contains 6 muffins. If Angelina purchases 4 packages, solve by multiplying 6×4, which equals 24 muffins.
12. **C.** If 9 branches are removed from each of 3 trees then we solve by multiplying $9 \times 3 = 27$.

13. **A.** (In this array, there are 4 rows with 7 squares in each row, which correctly represents the expression 4×7.)

14. The array in this question is made up of 3 rows of apples with 9 in each row. When written as a multiplication expression, it is 3×9 (9×3 is acceptable as well).

DIVISION

DEFINITIONS/VOCABULARY

Dividend: The number that is to be divided in a division problem (the number 18 in the problem $18 \div 3 = 6$).

Divisor: The number that divides the dividend (the number 3 in the problem $18 \div 3 = 6$).

Quotient: The number that results in a division problem (the number 6 in the problem $18 \div 3 = 6$).

Division is the process of separating a specified number of items into equal groups. For example, let's say you are given the number 15 and asked to split it into 5 equal groups. You could draw 5 circles and then separate 15 dots among those circles, being certain to make the number of dots in each circle equal. In this case, you will end up with 3 dots in each circle. This tells us that $15 \div 5 = 3$.

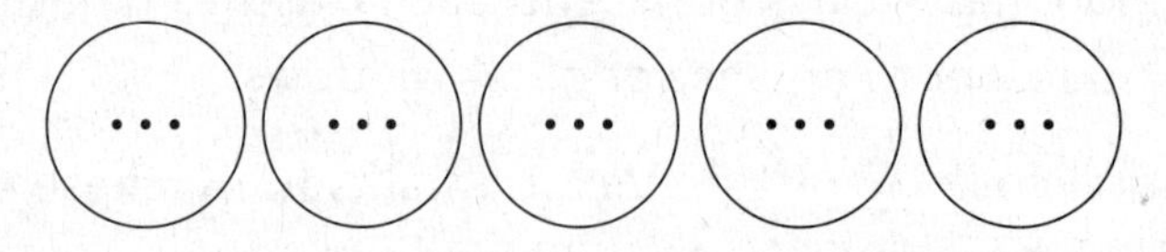

Division involves grouping, or equally breaking apart a given number into smaller groups. A division problem can be solved using an elementary method by using counters, or coins. The downside of this method is that it can take a very long time to count out coins, split them into equal groups, and hope that you did not drop one coin onto the floor or miscount originally. You also have to be certain you do not give any groups any more coins than the other groups have. The number of coins in each group most be equal. This can also be challenging.

Division should be thought of as the opposite of multiplication. If you are good at multiplication facts, you will probably be good at division facts. To solve smaller division problems, another strategy you can use is to read the problems backward as multiplication problems. For example, $24 \div 4 = n$ is a basic division problem. Since, multiplication is the opposite of division, we can read this problem backward as $n \times 4 = 24$. Whether you read the problem as $24 \div 4 = n$ or as $n \times 4 = 24$, the missing variable n is equal to 6.

Another way to compute a division problem is by skip counting. If you were given the problem $10 \div 2 = n$, you could solve this pretty easily with skip counting. How many times do you have to count by 2s until you reach 10? If you counted aloud or in your head, you would see that you need to count by 2s a total of 5 times in order to reach 10. Hence the answer is 5. This method really works with only very small division problems.

Skip counting becomes difficult if you are asked a question such as $49 \div 7 = n$ because counting by 7s can be confusing. In this case, you could use your knowledge of multiplication facts to solve for the missing variable. Something times 7 is equal to 49. What is the answer? The answer is 7. This tells us there are seven 7s in 49. Alternatively, you could take a piece of scratch paper, make 7 circles on it, and put a dot into each circle one at a time until you reach 49. Just make sure you have an equal number of dots in each circle. When finished, simply count how many dots are in each circle, and presto, you have your answer. Here's what your drawing might look like:

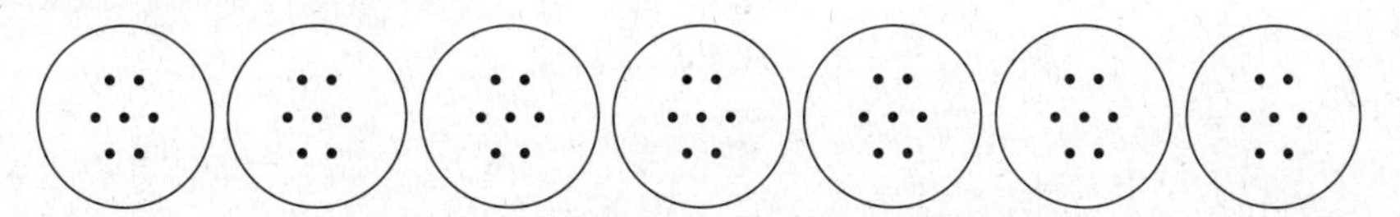

As you can imagine, this method also leaves a lot of room for error. Remember, though, that this method is an alternate strategy in case you have no other reasonable means to figure out the answer.

A third division strategy we can use is comparison bar modeling.

Example: James finds a box of crayons and markers in his desk drawer. He counts 32 crayons. There are 3 times as many crayons as markers. How many markers are in the box?

In this example, we know that two quantities are being compared. One quantity is a multiple of the other. We also know the larger quantity (the crayons). To find the smaller quantity, we can solve 32 ÷ 4 by using a comparison bar model:

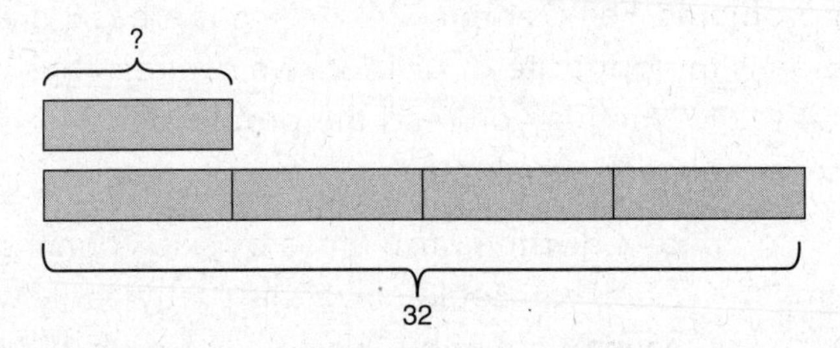

By using the division bar model, we can see that there is some number that is repeated 4 times to equal a quantity 32. That number is 8.

In third grade students must learn all of the division facts, similar to that of the multiplication facts, within 100. The process of long division begins in fourth grade where students will hand write division problems using a longer method because the division problems will use larger numbers. These problems may also include remainders where you cannot divide into equal groups. So now it is important for students to memorize those multiplication and division facts.

Division Exercises

1. What is 42 ÷ 6?
   - O A. 6
   - O B. 7
   - O C. 12
   - O D. 252

2. What is 25 ÷ 5?
   - O A. 5
   - O B. 125
   - O C. 3
   - O D. 4

3. Which is 63 ÷ 7?
   - O A. 461
   - O B. 8
   - O C. 7
   - O D. 9

4. What is $30 \div 5$?

- ○ A. 6
- ○ B. 15
- ○ C. 5
- ○ D. 2

5. What is $36 \div 9$?

- ○ A. 2
- ○ B. 8
- ○ C. 4
- ○ D. 7

6. Given the following division sentence: $32 \div b = 8$, which of the following multiplication sentences would help you solve it?

- ○ A. $8 \times 32 = b$
- ○ B. $b = 32 \times 8$
- ○ C. $32 \times 8 = b$
- ○ D. $b \times 8 = 32$

7. Nadia is organizing her desk organizer. She has 27 pencils in her desk. Nadia puts an equal number of pencils in each of the desk organizer's 3 slots. How many pencils did Nadia put in each slot?

- ○ A. 9
- ○ B. 8
- ○ C. 24
- ○ D. 7

8. Rocky purchased a bag of 12 apples. He gave 3 apples to each of his teachers. How many teachers did Rocky give apples to?

- ○ A. 36
- ○ B. 9
- ○ C. 6
- ○ D. 4

9. The Stevens family is packing up their 24 dinner plates to get ready to move. Each moving box holds 8 plates safely. How many boxes will the Stevens family need to safely pack their dinner plates?

- ○ A. 3
- ○ B. 2
- ○ C. 16
- ○ D. 14

10. There are 16 friends planning a trip to the local theme park. The friends will be driving to the theme park. If each car holds 4 people how many cars will the friends need if they are all going to the theme park?

- ○ A. 20
- ○ B. 12
- ○ C. 4
- ○ D. 5

11. There are 36 magazines in 4 equal stacks. How many magazines are in each stack?

- ○ A. 40
- ○ B. 9
- ○ C. 8
- ○ D. 32

12. The Dunesburg school district just purchased 27 new laptop computers for its classrooms. The school district has 3 buildings and the laptops must be split up equally among the 3 buildings. How many laptops will each building receive?

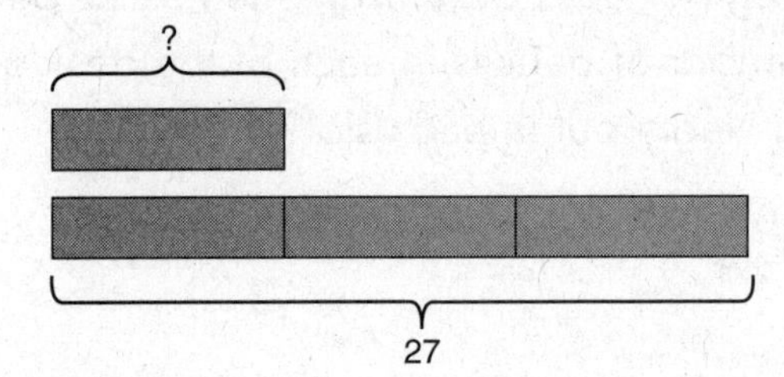

- ○ A. 30
- ○ B. 24
- ○ C. 9
- ○ D. 19

13. Thirty-six students are on Bus 66. They are on the way to school. Each row on the bus has 4 students. Write a number sentence to show how many rows the students fill.

Explain your answer: --

----

14. The third grade classes are planning a field trip to a farm. There are 72 students split equally among 8 chaperones. Write a number sentence to show how many students there will be in each group.

Explain your answer: --

Answers to Division Exercises

1. **B.** In the problem 42 ÷ 6, there are a total of seven 6s in the number 42.
2. **A.** In the problem 25 ÷ 5, there are a total of five 5s in the number 25.
3. **D.** In the problem 63 ÷ 7, there are a total of nine 7s in the number 63.
4. **A.** In the problem 30 ÷ 5, there are a total of six 5s in the number 30.
5. **C.** In the problem 36 ÷ 9, there are a total of four 9s in the number 36.
6. **D.** In the division sentence $32 \div b = 8$, we are missing one number. Remember, division is the opposite of multiplication. They are interrelated. Therefore, the correct answer is $b \times 8 = 32$. In both cases, $b = 4$.
7. **A.** If there are 27 pencils and 3 slots to put them in, we solve by computing 27 ÷ 3 = 9 pencils in each slot.
8. **D.** If there are 12 apples and Rocky hands out 3 to each of his teachers, that means he has 4 teachers because 12 ÷ 3 = 4.
9. **A.** If there are 24 dinner plates and each box safely holds 8, then we solve by computing 24 ÷ 8 = 3 boxes.
10. **C.** If there are 16 people and each car holds 4, then we solve by computing 16 ÷ 4 = 4 cars.
11. **B.** If there are 36 magazines split equally between 4 stacks, we can figure out how many are in each stack by dividing. There are 36 ÷ 4 = 9 magazines in each stack.
12. **C.** The bar model in this example is a great visual to help understand the mathematical process here. If there are 27 laptops to be distributed evenly to 3 schools, that means we are solving 27 ÷ 3. This is a basic division problem. You can think of it as a multiplication problem with some factor multiplying by 3 to equal a product of 27.
13. 36 ÷ 4 = 9. If there are 36 students altogether and 4 students in each row, we can divide to figure out how many rows are filled.

14. $72 \div 8 = 9$. If there are 72 students split equally among 8 chaperones, we can divide to find out how many students are in each group.

Multiplication and Division Fact Families

Division and multiplication are directly related to each other in the same way addition and subtraction are related. They are so related that you can create fact families of 4 problems (2 division and 2 multiplication) using just 3 numbers, which is exactly what you can do with addition and subtraction. In a fact triangle such as this, you would have two multiplication problems. Let's first review some vocabulary. In a multiplication problem resulting from a fact triangle, each equation has two factors and one product. The factors are the numbers being multiplied together, and the product is the result. Two division problems can also be created. Each consists of a dividend (the number that is to be divided in a division problem), a divisor (the number that divides the dividend), and a quotient (the answer). Shown graphically as a fact triangle the problem would look like this:

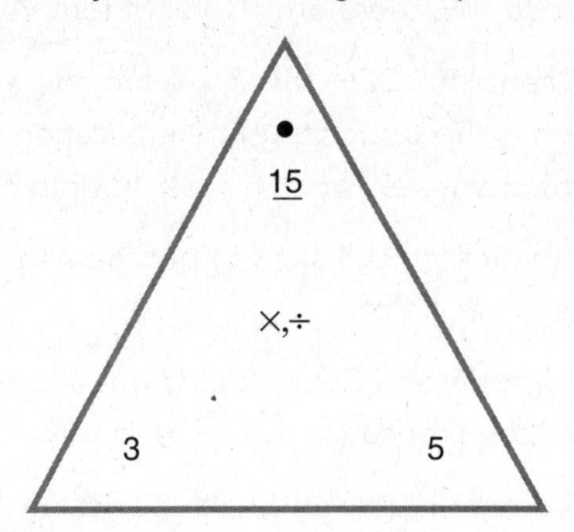

From this we can derive:

$3 \times 5 = 15$

$5 \times 3 = 15$

$15 \div 3 = 5$

$15 \div 5 = 3$

It is important to note that on fact triangles, the number on top has a dot just above it. For the two multiplication problems, this is the product (the result you get when you multiply the two other numbers). For the two division problems, this is the dividend (the number that is to be divided). In this type of fact triangle, any of the values may be missing, and it is your responsibility to determine the correct missing value, so it is important to understand the meaning of each value's location on the fact triangle.

On an exam, you may be asked to create a fact family for a missing variable. For example:

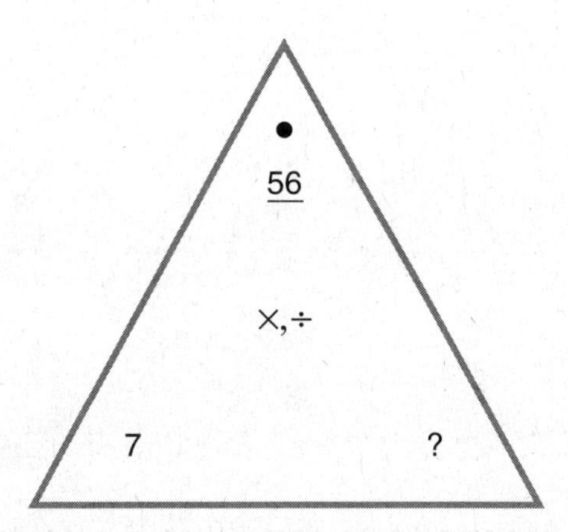

If you are shown the fact triangle above, you need to solve for the missing variable. On an exam, you may see the missing variable listed as a question mark (?), or a letter such as *x*, *r*, or *b*. You should be able to easily find the missing variable, and you can extend the solution by creating the four facts that go along with this triangle:

$\underline{56 \div 7 = ?}$

$\underline{56 \div ? = 7}$

$\underline{7 \times ? = 56}$

$\underline{? \times 7 = 56}$

If you know your multiplication facts well, you will realize the answer is 8. The division process here is taking the number 56, splitting it into 7 equal groups, and then recognizing that you would have exactly 8 in each group with none left over. Recalling your multiplication facts provides the best and quickest way to come up with a solution. You need to remember from this exercise that there is a direct relationship between multiplication and division. This relationship includes solving for a missing variable and creating a fact family.

Missing Variable and Division/Multiplication Fact Families Exercises

1. What is the value of the missing variable in the fact triangle shown below?

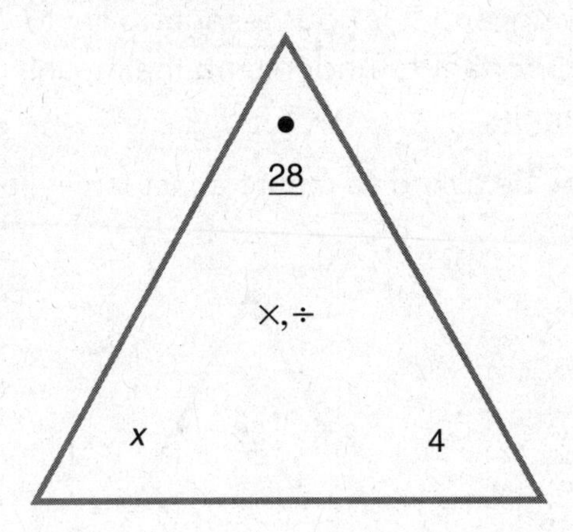

- ○ A. 3
- ○ B. 7
- ○ C. 4
- ○ D. 148

2. What is the value of the missing variable in the fact triangle shown below?

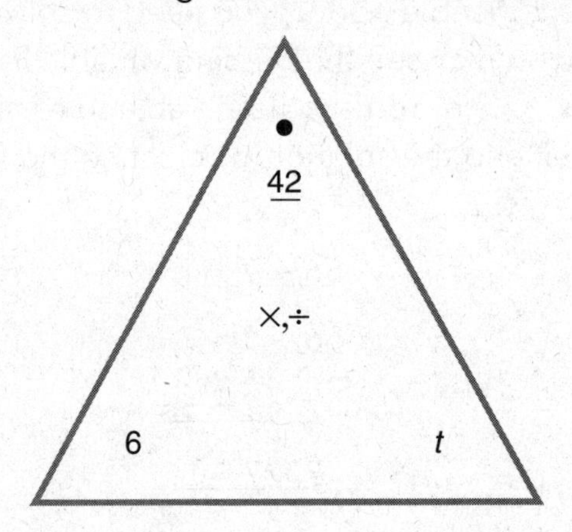

- ○ A. 7
- ○ B. 6
- ○ C. 4
- ○ D. 8

3. What is the value of the missing variable in the fact triangle shown below?

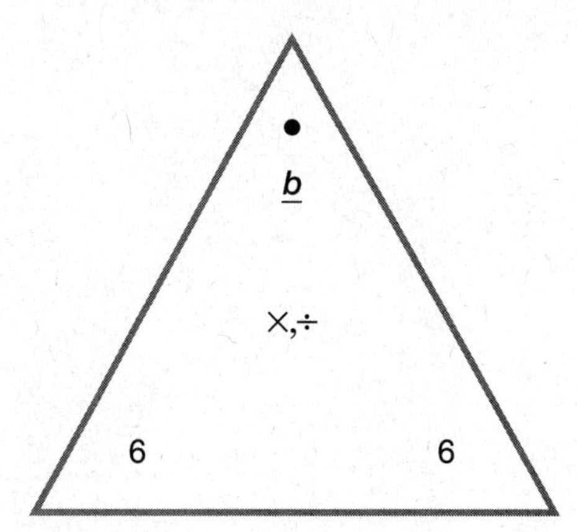

- A. 1
- B. 12
- C. 0
- D. 36

4. What is the value of the missing variable in the fact triangle shown below?

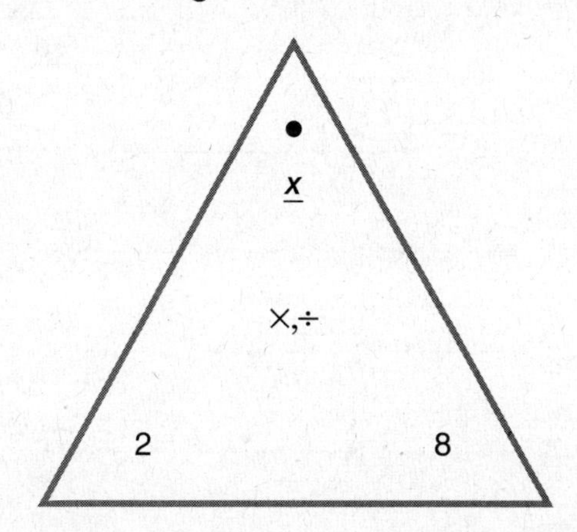

- A. 4
- B. 3
- C. 16
- D. 12

5. What is the value of the missing variable in the fact triangle shown below?

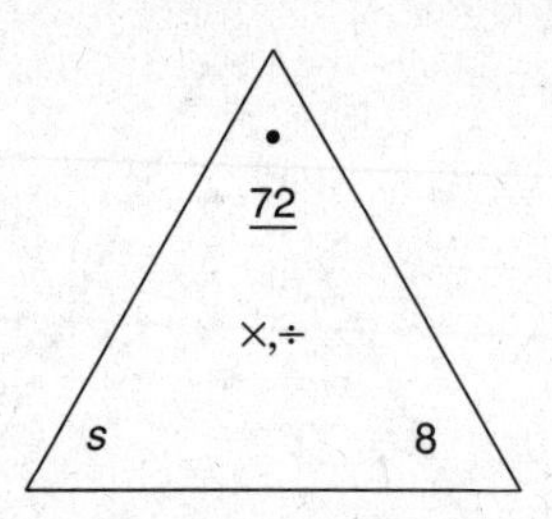

- ○ A. 9
- ○ B. 8
- ○ C. 11
- ○ D. 6

6. What is the fact family for the multiplication/division fact triangle shown below?

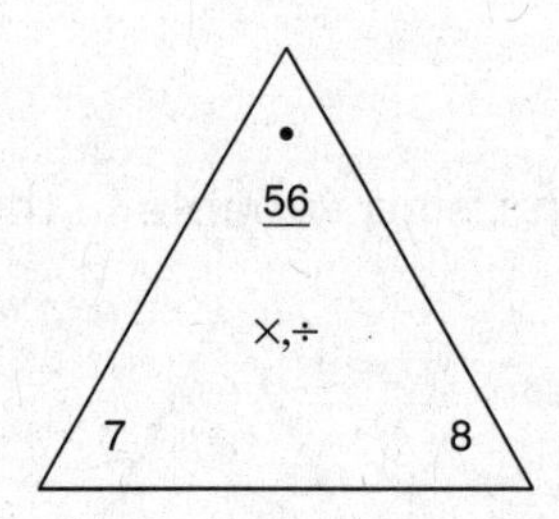

Answer: ________ × ________ = ________

________ × ________ = ________

________ ÷ ________ = ________

________ ÷ ________ = ________

7. What is the fact family for the multiplication/division fact triangle shown below?

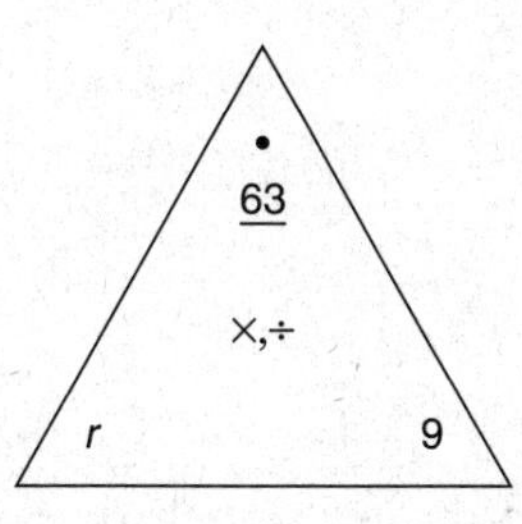

Answer: ________ × ________ = ________

________ × ________ = ________

________ ÷ ________ = ________

________ ÷ ________ = ________

8. What is the fact family for the multiplication/division fact triangle shown below?

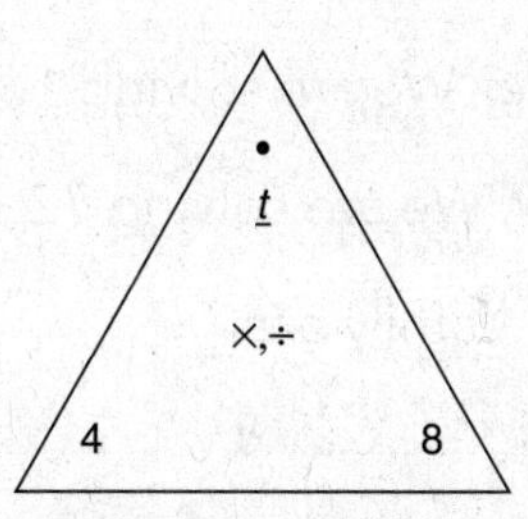

Answer: ________ × ________ = ________

________ × ________ = ________

________ ÷ ________ = ________

________ ÷ ________ = ________

9. Which two numbers can be placed in the blank spaces to make this number sentence true?

______ × ______ = 56

○ A. 7 and 8
○ B. 3 and 12
○ C. 7 and 7
○ D. 9 and 6

10. In the number sentence below, what does b equal?

$63 \div b = 7$

Answer: ____________________

Answers to Missing Variable and Division/Multiplication Fact Families Exercises

1. **B.** In this fact family, $x = 7$. We are solving $28 \div 4 = 7$, or alternately $7 \times 4 = 28$.
2. **A.** In this fact family, $t = 7$. We are solving $42 \div 6 = 7$, or alternately $7 \times 6 = 42$.
3. **D.** In this fact family, $b = 36$. We are solving $36 \div 6 = 6$, or alternately $6 \times 6 = 36$.
4. **C.** In this fact family, $x = 16$. We are solving $16 \div 2 = 8$, or alternately $2 \times 8 = 16$.
5. **A.** In this fact family, $s = 9$. We are solving $72 \div 8 = 9$, or alternately $8 \times 9 = 72$.
6. The four facts for this fact family are

   ____7____ × ____8____ = ____56____

   ____8____ × ____7____ = ____56____

   ____56____ ÷ ____7____ = ____8____

   ____56____ ÷ ____8____ = ____7____

7. Before listing the 4 facts, we must first solve for r. Since the product of 63 is given, and one factor is 9, we know that the other factor must be 7 since $9 \times 7 = 63$. Now we can proceed to list the four facts for this fact family:

   ____9____ × ____7____ = ____63____

   ____7____ × ____9____ = ____63____

   ____63____ ÷ ____7____ = ____9____

   ____63____ ÷ ____9____ = ____7____

8. Before listing the 4 facts, we must first solve for *t*. Since the two factors are given (4 and 8), we can simply multiply to find the missing value (we know this must be the product because it is notated with the dot). 4 × 8 is 32, which is our missing value. Now we can proceed to list the four facts for this fact family:

4	×	8	=	32
8	×	4	=	32
32	÷	4	=	8
32	÷	8	=	4

9. **A.** Only two factors given multiply to 56. That is 7 and 8.
10. $b = 9$. Because 63 divided by 9 equals 7, the variable *b* must be 9.

Single-Step Word Problems

Now let's apply some of the strategies we have covered. In previous fact practice sections you were faced with some word problems. The purpose was to get your mind thinking critically about what you were reading. Remember that mathematics doesn't always appear in the form of a number model. Countless real-world scenarios every day use mathematics in the form of word problems. You also probably know that word problems often sound more difficult than they are. At times, you might find it helpful to imagine that you and/or your friends/family are taking part in the word problem. If you make the problem personal, you will understand and solve the problem more easily. Overall, solving word problems requires critical thinking techniques so that you can identify important information and how it relates to the missing variable. You must read each problem slowly and carefully so you do not choose an incorrect operation to solve it.

Let's try a practice problem.

Example: Mrs. Nicosia just purchased 9 boxes of crayons to replace some broken ones in her classroom. Each box of crayons holds 8 crayons. How many crayons in all did she purchase?

First, let's identify the information given. Then we can look toward figuring out how to compute the missing variable. To organize our work we'll fill in a table:

Number of Boxes of Crayons	Number of Crayons per Box	Number of Crayons in All
9	8	?

Answer: ....
(unit)

Now that our information is organized and easily viewed, we can solve the problem. We can see that we'll have to use multiplication to find out how many crayons in all Mrs. Nicosia purchased. If each box has 8 crayons in it, we need to multiply that by the 9 boxes that were purchased. Another way to look at it would be to count by 8s nine times. You have multiple strategies at hand to solve this type of math fact, so feel free to use the one that works best for you. In the end, the correct answer in this problem is 72 crayons. (Remember to include the unit "crayons" in your answer.)

Single-Step Word Problems Exercises

1. Hilga has 24 books, in 4 equal piles. How many books are in each pile?

Number of Piles	Number of Books per Pile	Number of Books in All

Answer: ....
(unit)

Explain your answer: ....

....

....

2. Wally sees a full parking lot that has 6 rows of cars, and each row has 8 cars in it. How many cars does Wally see in this lot?

Number of Rows	Number of Cars per Row	Number of Cars in All

Answer: ....
(unit)

Explain your answer: ....

....

....

3. Jamie teaches gymnastics. She teaches 3 classes each day, 4 days a week. How many classes will Jamie have taught in one week?

Number of Days	Number of Classes per Day	Number of Classes in All

Answer: ....
(unit)

Explain your answer: ....

....

....

4. Jennifer bought tacos for her 7 friends and herself. She bought 2 tacos each for herself and each of her friends. How many tacos did she buy?

Number of People	Number of Tacos per Person	Number of Tacos in All

Answer: ....
(unit)

Explain your answer: --

----

5. Teddy's tenth grade class is going to the Bluewater Aquarium. The school has vans to get the students there. Each van holds 6 students, and there are 24 students in the class. How many vans will be needed to drive the students to the aquarium?

Number of Vans	Number of Students per Van	Number of Students in All

Answer: ....
(unit)

Explain your answer: --

Answers to Single-Step Word Problems Exercises

1.

Number of Piles	Number of Books per Pile	Number of Books in All
4	?	24

Answer: ....6.......... books per pile
(unit)

If there are 24 books altogether, and they are split among 4 piles, then this is a division problem: $24 \div 4 = 6$. If you do not have the multiplication facts memorized, you can use some of the methods discussed in this chapter to find the answer of 6. Examples would be building an array or sorting into groups.

2.

Number of Rows	Number of Cars per Row	Number of Cars in All
6	8	?

Answer: ....48.......... cars in all
(unit)

In this problem, we have 6 rows with 8 cars in each row. This is a basic multiplication problem as we are counting by 8s six times. (Alternately, we could also count by 6s eight times.) If you do not know the answer to 6×8 you could build a 6×8 array and count the elements in the array to arrive at an answer of 48.

3.

Number of Days	Number of Classes per Day	Number of Classes in All
4	3	?

Answer: ....12.......... classes in all
(unit)

In this problem Jamie is teaching 3 classes per day, and she is doing this 4 times each week. This is a basic multiplication problem as we are counting by 3s four times. (Alternately, we could also count by 4s three times.) If you do not know the answer for 4×3, you could build a 4×3 array and count the elements in the array to arrive at an answer of 12.

4.

Number of People	Number of Tacos per Person	Number of Tacos in All
8	2	?

Answer: ....16......................tacos in all
(unit)

This problem is a little tricky because you must remember to include Jennifer. She is also going to eat tacos. Including Jennifer, there are 8 people, and each person is going to eat 2 tacos. This is a multiplication problem. You can count by 8s two times, or count by 2s 8 times. In either case, you are solving $8 \times 2 = 16$.

5.

Number of Vans	Number of Students per Van	Number of Students in All
?	6	24

Answer: ....4......................vans
(unit)

This is a division problem. If we know 24 students must be split into groups of 6 to fit in vans, then we are solving $24 \div 6$. Our lesson on fact families taught us that this is really asking what number times 6 equals 24. Thinking through your 6s facts, you should come up with an answer of 4. If you did not memorize your facts yet, you could take the 24 students and split them into equal groups of 6. Then you would see that there are 4 groups (vans) in total.

Two-Step Word Problems

Two-step word problems are slightly more advanced than the single-step problems previously discussed. The difference is implied in the name. Two-step word problems require you to do multiple mathematical operations before you can begin solving the problem. For example, you may have to add some numbers or divide two numbers before you can move on to another operation to get your final answer. The level of critical thinking is more advanced because you must determine what combination of the four operations (×, ÷, +, −) you need to use. Let's take a look at an example.

Example: Mr. Jenkins travels a good distance to work every day. In his work commute, Mr. Jenkins' car uses 2 gallons of gas per day. If he works 5 days per week, how many gallons of gas would he use per month (4 weeks) traveling back and forth to work?

The answer to this problem requires two mathematical steps. First, if the car uses 2 gallons per day, we need to figure out how many gallons it uses per week. The problem states that Mr. Jenkins works 5 days per week, so we multiply 2×5 and get 10. Each week his car uses 10 gallons of gas as he travels to and from work. Now that we know the weekly total for gas, we can multiply this by 4 since there are 4 weeks in the month. $10 \times 4 = 40$. So each month, the car uses 40 gallons of gas due to Mr. Jenkins' work. Arithmetically, this problem would be written as $2 \times 5 \times 4 = 40$. Because we multiplied two times in this example, we can use the associative property of multiplication as a strategy for solving. According to the associative property of multiplication, there are two ways to solve this problem. It can either be solved as $(2 \times 5) \times 4 = 40$ or $2 \times (5 \times 4) = 40$. In each example you would solve what's in the parentheses first and then multiply it by the remaining factor to get the product. In essence, the property states that you can multiply any two factors together first and then multiply by the third factor to get the final answer. Note: The problem required multiplication in both steps, but it should be noted that two-step word problems may involve any combination of the 4 mathematical operations.

Two-Step Word Problems Exercises

1. Ron is buying some fresh fruit. He bought 6 oranges, 3 grapefruit, 11 lemons, and 20 limes. If each piece of fruit costs $0.10, how much did the fruit cost in all?

   Explain your answer: --

   ----

2. Marilyn likes to walk her grandson Mark, to the park for a stroll each day. Mark drinks one bottle of water every hour while at the park. If they will be at the park for 2 hours, how many bottles of water will Marilyn need to purchase each week (7 days) to take to the park?

Explain your answer: --

----

3. Carly babysits for the next door neighbor. Last month, she babysat for 7 nights and sat the following number of hours: 2 hours, 3 hours, 2 hours, 4 hours, 5 hours, 3 hours, and 1 hour. If she is paid $6.00 per hour, how much money did she make babysitting for the next door neighbor last month?

Explain your answer: --

----

4. Tom cuts lawns for a little cash. If he cuts three lawns per week, and works 3 hours on one, 2 hours on a second one, and 4 hours on a third one, how many hours will he have worked in a month (4 weeks)?

Explain your answer: --

----

5. Yanni went to the video store and bought 6 DVDs, 11 CDs, and 7 Blu Ray discs. When he got home, he realized he already had two of the DVDs he just purchased. So as to not waste money, Yanni went back to the store and returned those two DVDs. After returning those two DVDs, how many items did Yanni end up purchasing?

Explain your answer: --

Answers to Two-Step Word Problems Exercises

1. \$4.00. If we count all the fruit that Ron purchased, we get 40 (6 + 3 + 11 + 20 = 40). Now that we have the total number of pieces of fruit, we need to multiply it by \$0.10 to determine the total cost. 40 × \$0.10 is \$4.00. Arithmetically this problem would be written out as (6 + 3 + 11 + 20) × \$0.10 = \$4.00.
2. 14 bottles. If Mark drinks a bottle per hour at the park, and he's there for 2 hours each day, we need to multiply 1 × 2 to determine how many bottles are needed each day at the park. Since 2 bottles are needed each day, we must then multiply the 2 bottles of water Mark needs each day by 7 (the number of days they go to the park). 2 × 7 = 14. So Marilyn needs to purchase 14 bottles of water to meet Mark's needs for water each week at the park. Arithmetically this problem would be written out as 1 × 2 × 7 = 14.
3. \$120.00. First, we need to determine the total number of hours Carly babysat. To do this, we add up all of her hours: 2 + 3 + 2 + 4 + 5 + 3 + 1 = 20. Now that we know she babysat for a total of 20 hours last month, we can multiply 20 hours by how much money she is paid each hour (\$6.00). 20 × \$6.00 = \$120.00. So last month Carly made \$120.00 babysitting for her neighbor. Arithmetically this problem would be written out as (2 + 3 + 2 + 4 + 5 + 3 + 1) × \$6.00 = \$120.00.

4. 36 hours. We need to first figure out how many hours Tom mows each week. The problem tells us he mows three lawns per week. To get the total, we need to add the number of hours it takes to mow each lawn each week. $3 + 2 + 4 = 9$. So each week Tom spends 9 hours mowing lawns. However, the problem is asking how many hours he works in a month (4 weeks), so we must multiply 9 (hours per week) × 4 (weeks) = 36 hours. Arithmetically, this problem would be written out as $(3 + 2 + 4) \times 4 = 36$.

5. 22 items. When Yanni went to the store, he originally purchased a total of 24 items ($6 + 11 + 7 = 24$). When he got home and realized he already had two of the items, he returned them to the store. So Yanni now has two less items than he originally did when he left the store. $24 - 2 = 22$. Yanni ended up with 22 new items. Arithmetically, this problem would be written out as $(6 + 11 + 7) - 2 = 22$.

Patterns

Patterns are all around us. Many of the nursery rhymes that you are familiar with describe patterns.

Hickory, dickory, dock.
A mouse ran up the clock.
The clock struck one;
And down he run.
Hickory, dickory dock.

It has been suggested that the words hickory, dickory, and dock stand for "eight", "nine", and "ten," respectively. In the schoolyard verse for choosing sides, "eeny, meeny, miny, mo" most likely stand for "one, two, three, four."

Do you remember this familiar nursery rhyme?

As I was going to St. Ives,
I met a man with seven wives,
And every wife had seven sacks,
And every sack had seven cats,
And every cat had seven kits.
Kits, cats, sacks, wives,
How many were going to St. Ives.

It suggests a multiplication problem: $7 \times 7 \times 7 \times 7$.

Let's Explore a Pattern.

Activity: Get out a piece of paper. Write 7 up at the top. Add 5 to the 7 and put the result below the 7. Add 5 to the result and put the result below. Continue to do this until you get to around 120. Now, look at the sequence of numbers.

Do you notice any patterns? Are there any patterns in the units place? In the tens place? Add two successive numbers. Do you see a pattern?

Mathematics is full of patterns. Some patterns are easy to see, such as the even numbers (2, 4, 6, 8, etc.). Some are not so easy to see, such as the Fibonnacci sequence (1, 1, 2, 3, 5, 8, 13, 21, etc.). In case you didn't see the pattern in the Fibonnacci sequence, add the first two numbers together. Now add the next two numbers. Each number in the sequence is found by adding the previous two numbers together. Mathematics is considered to be the search for patterns in our world. A formula, such as the area of a rectangle ($A = L \times W$), or area is length times width, is nothing more than a pattern for finding how big a rectangle is. Using patterns, we can find out how big a wall is, or how big a room is (volume). Using patterns, we can predict (to the second) when the sun will rise and when it will set.

Number arrays (as discussed earlier) can have patterns. The array of numbers:

2, 4, 6, 8, 10, 12, etc.

is probably familiar to you. It is the even numbers.

Now, look at this pattern:

1, 3, 5, 7, ...

What would the next three numbers in this array be? The next three numbers in this array are 9, 11, and 13 because these are the odd numbers.

You can count down and still find patterns in numbers. Look at this array:

55, 51, 47, 43, 39, ...

What would the next three numbers be? The next number would be 35, 31, and 27 because you are counting down by 4s.

Similarly, patterns can be represented in a table, as shown below. In this type of representation, you have three components: the rule, the IN number, and the OUT number. In the example below, we are given the rule of × 2. For those IN numbers we have, we can simply apply the rule and obtain the OUT number. So 6 × 2 = 12, 7 × 2 = 14.

RULE
× 2

IN	OUT
6	?
7	?
?	18
?	6
4	?

So 6 × 2 = 12, 7 × 2 = 14, and 4 × 2 = 8. For those locations where we are missing the IN number, we must consider what number × 2 equals the OUT number. Alternately, we could think of this as a division problem: the OUT number divided by the rule.

RULE
× 2

IN	OUT
6	12
7	14
9	18
3	6
4	8

The concept of patterns can be further extended in this type of table by giving all variables in the table and requiring you to determine the rule. In an example such as this, remember to compare multiple rows of numbers to be certain the rule is what it appears to be.

RULE
?

IN	OUT
41	47
101	107
3	9
14	20

Here, we take one of the IN numbers and compare it to the OUT number. If we compare 41 to 47, we can see it increased by 6, so the rule should be +6. However, to be certain, always check another row. If we move down and try the 3 and 9, we can see that the rule of +6 does in fact work.

Patterns Exercises

1. What are the next four numbers in the number array: 5, 8, 11, 14, ...

------------ ------------ ------------ ------------

Explain the pattern.

----

2. What are the next three numbers in the number array: 256, 251, 245, 238, ...

------------ ------------ ------------

Explain the pattern.

----

3. Complete the table below and apply the rule to make your own IN and OUT combination in the bottom row of the table.

RULE
× 6

IN	OUT
6	
4	
	30
	6

4. Complete the table below and apply the rule to make your own IN and OUT combination in the bottom row of the table.

RULE
× 4

IN	OUT
	28
	4
9	
0	

5. Complete the table below by finding the value for the rule, then apply the rule to make your own IN and OUT combination in the bottom row of the table.

RULE

IN	OUT
3	6
9	18
22	44
14	28

6. Complete the table below and apply the rule to make your own IN and OUT combination in the bottom row of the table.

RULE
÷ 2

IN	OUT
6	
	8
	5
20	

7. Complete the table below and apply the rule to make your own IN and OUT combination in the bottom row of the table.

RULE
+ 4

IN	OUT
	7
28	
	47
	61

8. Jason and Beverly came up with two math patterns. Each pattern has five numbers. Jason's pattern is: 88, 82, 76, 70, 64.

Beverly's pattern increases by the same number that Jason's decreases, and it also starts at the same number as Jason's.

What are the five numbers in Beverly's pattern?

Explain your answer:

---

9. Mark mows lawns every Saturday to earn some extra cash. Last Saturday he mowed six lawns and charged $15.00 for each of them. Write a pattern that shows how much money Mark had earned by the time he finished his sixth lawn.

Explain your answer:

---

10. Jeremy's parents pay him an allowance of $5.00 per week for doing chores around the house.

5, 10, 15, 20, 25, 30, 35, 40

The pattern above shows how much allowance Jeremy earned in 8 weeks.

How much money did Jeremy earn after 6 weeks?

Explain your answer:

Answers to Patterns Exercises:

1. The next four numbers in the array would be 17, 20, 23, and 26.

   The pattern is that the next number in the array is increased by 3.

2. The next three numbers in the array would be 230, 221, and 211.

   The pattern is that the next number in the array is decreased by one more each time. For example, the difference between the first and the second number is 5; the difference between the second and third numbers is 6; the difference between the third and fourth numbers is 7, and so forth.

3. In this example we are given the rule of ×6. We can simply apply this rule to our given IN numbers to obtain the OUT numbers. 6 × 6 = 36 and 4 × 6 = 24. For the missing IN numbers, we need to think of it as __ × 6 = 30 or we can think of it in terms of division: 30 ÷ 6 = __. In either case, the answer is 5. We can also apply it to the 6. Think of it as 6 ÷ 6 = __, or __ × 6 = 6. In both cases the answer is 1.

RULE
× 6

IN	OUT
6	36
4	24
5	30
1	6

4. In this example we are given the rule of ×4. We can simply apply this rule to our given IN numbers to obtain the OUT numbers. 9 × 4 = 36 and 0 × 4 = 0. For the missing IN numbers, we need to think of it as __ × 4 = 28, or we can think of it in terms of division: 28 ÷ 4 = __. In either case, the answer is 7. We can also apply it to the 4. Think of it as 4 ÷ 4 = __, or __ × 4 = 4. In both cases the answer is 1.

RULE
× 4

IN	OUT
7	28
1	4
9	36
0	0

5. In this example, we are given all of the IN and OUT values and we are asked to determine the rule. To determine the rule, we need to pick a row and figure out what's happening. Let's look at the 3 and the 6. At first glance you might think + 3. However, *always* check another row to be certain. Let's look at the 22 and the 44. We can quickly see that the rule is not + 3 because it does not work on this row. Instead, the rule is *double* (or think of it as × 2).

RULE
Double

IN	OUT
3	6
9	18
22	44
14	28

6. In this example, we are given the rule of ÷ 2. We can simply apply this rule to our given IN numbers to obtain the OUT numbers. 6 ÷ 2 = 3 and 20 ÷ 2 = 10. For the missing IN number 8, we need to think in terms of either division __ ÷ 2 = 8 or multiplication 8 × 2 = __. In either case, the answer is 16. Similarly, we can apply it to the 5. Think of the OUT number as __ ÷ 2 = 5, or 5 × 2 = __. In both cases, the answer is 10.

RULE
÷ 2

IN	OUT
6	3
16	8
10	5
20	10

7. In this example, we are given the rule of +4. We can simply apply this rule to our given IN number to obtain the OUT number. 28 + 4 = 32. For the missing IN numbers, we need to think of it as __ + 4 = 7 or we can think of it in reverse using subtraction: 7 − 4 = __. If we apply this same thinking to the 47 we get 43: __ + 4 = 47, or 47 − 4 = __. Finally, for the 61, we get an answer of 57: __ + 4 = 61 or 61 − 4 = __.

RULE
+ 4

IN	OUT
3	7
28	32
43	47
57	61

8. We can tell by looking at Jason's pattern that it decreases by 6 (88 − 82 = 6). If Beverly's five-number pattern *increases* by the same number (6) and starts at the same number (88), then her pattern would have to be 88, 94, 100, 106, 112. Here the pattern is +6 instead of Jason's −6.

9. If Mark mowed six lawns and charged $15.00 each for them, then by the time he finished his sixth lawn he would have a pattern of earnings that looks like this:

   15, 30, 45, 60, 75, 90

   Each number in the pattern represents the accumulated total he earned. So by the time he finished his sixth lawn he had earned a total of $90.00.

10. The pattern given in this problem shows Jeremy's allowance earnings for a period of 8 weeks. However, the question is only asking for his earnings after 6 weeks. So we can take a look at the pattern and look specifically for the sixth number:

    5, 10, 15, 20, 25, 30, 35, 40

    Given this pattern, the sixth number is 30. This tells us that Jeremy had earned a total of $30 in allowance after the sixth week.

Number and Operations in Base Ten

CHAPTER 4

Numbers are all around us. We use and see numbers every day. Numbers can be used to represent given values, but they can also be used in mathematical operations. In the grocery store alone, numbers are used in many different ways: weights of items, costs of items, aisles in the store, quantity in a package, and many more. Also, in the grocery store it can become quite difficult to keep track of exactly how much your groceries are going to cost. We can estimate how much the groceries will cost by rounding numbers to make mental math computations easier. To understand the process of rounding, we must first discuss place value concepts.

Whole Number Place Value

DEFINITIONS/VOCABULARY

Place Value: every digit in a number has what is called a **place value**. The place value of a digit is determined by its location in relation to the decimal point.

In this chapter, we will look at place values of whole numbers to the nearest 10 or 100.

5,461

The number above is read as five thousand four hundred sixty-one. Note that the word "and" was not used anywhere in this extended notation. The word "and" is commonly said out loud when repeating a long number such as this. In fact, it is often inserted into the sentence in place of the comma. However, the word "and" more formally refers to a decimal position in a number, not a comma.

In the number 5,461 each digit has a specific place value that is determined by its overall position relative to the decimal. The place values are listed as follows:

Number	5,	4	6	1
Place Value	Thousands	Hundreds	Tens	Ones

The ones position is farthest to the right. A decimal usually comes after that position. If there are no digits to the right of the decimal, the decimal is generally not written (as in this example). We assume that the decimal is located at the end of the number in the ones position. When we read place values from the decimal (going to the left), we start with ones and move to tens, hundreds, thousands. When we discuss decimals the same rule will be used, but this time we will be going to the right of the decimal. In the whole number 5,461 listed above, the place value of each digit is listed underneath it. The place value is permanent, or fixed. The place value of each digit gives that specific digit a certain value based on its location. For example, in the number 5,461 the 4 is in the hundreds position. Therefore, the 4 is worth 400 because there are 4 of them in the hundreds position. To prove this you could count by hundreds four times. At the end you would have an answer of 400. The 5 is in the thousands position, so this digit would be worth 5,000 because 5 thousands are a total of 5,000. We will get more into place value rounding in the next section. For now, try a few practice problems on place value.

Whole Number Place Value Exercises

1. In the number 3,091, which digit is located in the hundreds place?

O A. 3
O B. 0
O C. 9
O D. 1

2. In the number 3,478, which digit is located in the ones position?

O A. 4
O B. 3
O C. 8
O D. 7

3. In the number 8,562, which digit is located in the thousands position?

O A. 5
O B. 2
O C. 6
O D. 8

4. In the number 1,037, which digit is located in the thousands position?

O A. 1
O B. 7
O C. 3
O D. 0

5. In the number 9,471, which digit is located in the hundreds position?

○ A. 7
○ B. 9
○ C. 1
○ D. 4

6. Select all the numbers that have the same digit in the hundreds place.

☐ A. 7,421
☐ B. 974
☐ C. 1,048
☐ D. 403

Answers to Whole Number Place Value Exercises

1. **B.** The 0 is the third digit from the right, which is the hundreds position.
2. **C.** The 8 is the first digit from the right, which is the ones position.
3. **D.** The 8 is the fourth digit from the right, which is the thousands position.
4. **A.** The 1 is the fourth digit from the right, which is the thousands position.
5. **D.** The 4 is the third digit from the right, which is the hundreds position.
6. **A.** and **D.** The question wants us to find all the numbers that have the same digit in the hundreds place. 7,421 has a 4 in the hundreds place; 974 has a 9 in the hundreds place; 1,048 has a 0 in the hundreds place; and 403 has a 4 in the hundreds place. Therefore, both 7,421 and 403 are the answers.

Rounding

DEFINITIONS/VOCABULARY

Rounding: the process in which we replace a number (or digit) with another number to make it easier to estimate a math problem using mental math.

Overall the purpose of rounding is to give us a number that is easy to work with when we have to use mental math. Rounding can be useful in many situations, particularly, it is very helpful in the supermarket. When shopping, you generally do not have a calculator handy to compute the cost of the items in your grocery cart. If you use mental math and rounding, it is possible to keep track of costs (as you shop.)

The rounding process is quite simple. To round, you must follow a few basic rules.

When rounding a number, if we are told to round it to the nearest hundred, we must look at the digit in the hundreds position. Then we need to look at the digit to the right. If the digit to the right is 5 or greater, we will bump up the number in the hundreds position and change the digits to the right to zeros. If the digit is less than 5, we will leave the number in the hundreds position the same and change the digits to the right to zeros.

For example, in the number 871, the 8 is in the hundreds position. If we were asked to round 871 to the nearest hundred, we would locate the 8, which is in the hundreds position, and then look to its right. The number to the right is 7. Seven is greater than 5 so according to the rounding rule, the 8 must be bumped up to 9. All the digits to the right (in the tens and ones places) become 0. So 871, rounded to the nearest hundred, will be 900. This makes sense because 871 is closer to 900 than it is to 800. As proof, it is 71 away from 800 (871 – 800 = 71) yet only 29 away from 900 (900 – 871 = 29).

Likewise, if we were asked to round the number 6,435 to the nearest thousand, we would follow the same steps. Looking at the number, we can see that we have a 6 in the thousands position. To the right of it is a 4. Four is less than 5, so applying the rounding rule we must leave the 6 alone and change all the numbers to the right into zeros. So 6,435 rounded to the nearest thousand would be 6,000 (6,435 is closer to 6,000 than it is to 7,000).

If we have a larger number, for example 42,361, we may be asked to round it to the hundreds position. This problem is still like the last two examples. The only difference is that there are extra digits to the left of the digit being rounded. So let's take a look. If we want to round 42,361 to the nearest hundred, we would point out the digit in the hundreds position, which is the 3. Looking to the right of it, we see a 6. Six is greater than 5, so the rule says to bump up the digit in the hundreds and make everything to the right zeroes. The digits to the left of the 3 stay as they are. So 42,361 rounded to the nearest hundred would be 42,400.

Rounding Exercises

1. What is 587 rounded to the nearest hundred?

○ A. 500
○ B. 580
○ C. 600
○ D. 590

2. What is 3,971 rounded to the nearest thousand?

- O A. 3,980
- O B. 3,900
- O C. 4,100
- O D. 4,000

3. Which response below shows a digit that is rounded to the hundreds position?

- O A. 9,640
- O B. 32,800
- O C. 5,222
- O D. 8

4. Which response below shows a digit that is rounded to the thousands position?

- O A. 76,000
- O B. 87,901
- O C. 14,814
- O D. 5,400

5. What is 3,971 rounded to the nearest thousand?

- O A. 3,900
- O B. 3,000
- O C. 3,970
- O D. 4,000

6. What number, if rounded to the tens, would have an answer of 80?

- O A. 86
- O B. 74
- O C. 84
- O D. 70

7. What number, if rounded to the hundreds, would have an answer of 900?

- O A. 944
- O B. 849
- O C. 951
- O D. 989

8. What number, if rounded to the hundreds, would have a 7 in the hundreds position?

- ○ A. 1,600
- ○ B. 1,632
- ○ C. 7,789
- ○ D. 1,656

9. In the number 4,901, what is the value of the 9?

- ○ A. 90
- ○ B. 900
- ○ C. 9
- ○ D. 9,000

10. Which two tens is the number 47 between?

- ○ A. 10 and 20
- ○ B. 20 and 30
- ○ C. 30 and 40
- ○ D. 40 and 50

11. Select all the numbers that round to the same value when rounded to the nearest hundred.

- ☐ A. 714
- ☐ B. 688
- ☐ C. 751
- ☐ D. 643

12. Thinking about the nearest ten, 614 is closer to ______ than it is to ______?

600 610 620

Answer: --

13. Estimate the difference by rounding each number to the nearest hundred.

781 – 314

Explain your answer: --

----

14. Mrs. Massimi and her family drive 391 miles to a sunny vacation spot. After one week, they head back home. This time they take a different route that ends up being 421 miles. Estimate to the nearest hundred how many miles they drove round-trip.

Explain your answer: --

----

15. Ronald wants to round the number 247 to the nearest ten. What is the answer?

Explain your answer: --

Answers to Rounding Exercises

1. **C.** 587 is closer to 600 than it is to 500, if you were rounding to the nearest hundred.
2. **D.** 3,971 is closer to 4,000 than it is to 3,000, if you were rounding to the nearest thousand.
3. **B.** 32,800 is the only number that is rounded to the hundreds position. In this number, the 8 is in the hundreds position and every other digit to the right is 0.
4. **A.** 76,000 is the only number that is rounded to the hundreds position. In this number, the 6 is in the thousands position and every other digit to the right is 0.
5. **D.** 3,971 is closer to 4,000 than it is to 3,000 if you were rounding to the nearest thousand.
6. **C.** Rounding to the tens: 86 = 90, 74 = 70, 84 = 80, and 70 = 70. Since the question is asking which one would be 80 when rounded to the tens, we can see the answer is C.
7. **A.** Rounding to the hundreds: 944 = 900, 849 = 800, 951 = 1,000, and 989 = 1,000. Since the question is asking which one would be 900 when rounded to the hundreds, we can see the answer is A.

8. **D.** 1,600 is already rounded to the hundreds; 1,632 would be 1,600; 7,789 would round to 7,800; and 1,656 would round to 1,700. Since the problem asks which number rounded to the hundreds has a 7 in the hundreds position, we can see the answer is D.
9. **B.** In the number 4,901, the 9 is in the hundreds position. Therefore, the value of this digit is 900.
10. **D.** If we have the number 47, the nearest ten going down would be 40. That is 7 less than where we started. The nearest ten going up would be 50. That is 3 more than where we started.
11. **A** and **B.** 714 rounded to the nearest hundred is 700; 688 rounded to the nearest hundred is 700; 751 rounded to the nearest hundred is 800; 643 rounded to the nearest hundred is 600. Looking at the rounded answers, we see only two that match (both are 700). Therefore, the answer is A and B (714 and 688).
12. 614 is closer to 610 than it is to 620. 614 is 4 away from 610, and it is 6 away from 620. Because 4 is less than 6, it is closer to 610. If you mark the location of 614 on the number line, you can see that it is closer to 610 than it is to 620.
13. 500. 781 rounded to the nearest hundred is 800; 314 rounded to the nearest hundred is 300; 800 – 300 = 500.
14. 800. 391 rounded to the nearest hundred is 400; 421 rounded to the nearest hundred is 400 as well; 400 + 400 = 800 miles.
15. 250. 247 rounded to the nearest ten is 250. 247 is only 3 away from 250 (250 – 247) and 7 away from 240 (247 – 240). Since it is closer to 250, that is our answer.

Addition

DEFINITIONS/VOCABULARY

The **Addends**: the numbers being added to reach the sum.

Sum: the amount you obtain by adding the addends together.

The math fact 9 + 8 = 17 is comprised of two addends and a sum.

In this example, 9 and 8 are addends, and 17 is the sum.

Addition is the arithmetic process of combining numbers. We use addition in many tasks throughout a normal day. Many times, we may not recognize an addition problem as a math problem. In fact, it may be something you do regularly but do not realize you are using math to do it. You probably use mental math to compute many smaller addition problems. You may also estimate by way of rounding as previously described. However, you do need to use paper and pencil or a calculator to solve more complex problems.

Smaller problems such as 9 + 8 (also known as addition facts, because both numbers being added are only one digit) can be done using mental math. Using mental math, you should come up with an answer of 17 for 9 + 8. This is a basic concept that should already be mastered by the time you reach Grade 3. You should have already memorized these basic facts to make the larger problems easier to deal with.

As the problems grow larger, it will also become necessary to use paper and pencil to find the answer. Students in the third grade should be able to add and subtract up to sums of 1,000.

Let's look at adding a two-digit number to another two-digit number such as 54 + 67. One of the keys to solving addition and also subtraction problems is rewriting them vertically (up and down). 54 + 67 is written horizontally (side by side) and is difficult to solve because the place values are not lined up on top of each other. Rewriting the problem and lining up the place value columns makes this problem much easier to solve.

$$\begin{array}{r} 5\ 4 \\ +\ 6\ 7 \\ \hline \end{array}$$

In any addition problem, you must always start adding on the right-hand side. So in our example, the first two digits we add are the 4 and the 7. This yields an answer of 11.

Underneath the 7, we write the first part of our answer, a 1. The 1 represents the digit in the ones position in the number 11. The other 1 (from the tens position) is carried over on top of the 5 and added to it.

$$\begin{array}{r} {\scriptstyle 1}\ \ \ \\ 5\ 4 \\ +\ 6\ 7 \\ \hline 1 \end{array}$$

Now you are ready to begin adding in the tens column. Here we have the 1 we carried, the 5, and the 6. This gives us a total of 12. Normally we would write the 2 underneath the 6, and carry the 1. However, there is nowhere to carry the 1 to.

There are no other digits to the left. If you pretend to carry the 1, you will see that it lands in the hundreds position, which is correct, even though there are no other digits to add to it. So you can either do this, or just bring it down to the bottom for a final answer of 121.

$$\begin{array}{r} {\scriptstyle 1} \\ 54 \\ +\ 67 \\ \hline 121 \end{array}$$

This process remains the same no matter how many digits you are adding. It could be 9,581 + 372. Again, you would rewrite the problem vertically (9,581 on top of the 372). Notice that the place values of the two numbers are lined up properly when we rewrite it. Everything is pushed as far to the right as possible.

$$\begin{array}{r} 9,581 \\ +\ \ \ 372 \\ \hline \end{array}$$

Now you can begin adding the digits starting in the ones position (all the way to the right), and moving over to the left until there is nothing more to add.

Be absolutely certain that when you rewrite the problem you line everything up on the right side.

$$\begin{array}{r} {\scriptstyle 1} \\ 9,581 \\ +\ \ \ 372 \\ \hline 9,953 \end{array}$$

After successfully adding all of the place value columns, we can see that our final answer works out to 9,953.

Let's delve a little further into the process of addition by looking at it through a word problem. Remember that in word problems you need to thoroughly read the problem before you can decide how to solve it.

Example: J. Miller Elementary School was holding a talent show. Students had to sign up in order to participate. At the end of the sign-up period, there were 41 boys and 11 more girls than boys. How many girls signed up for the talent show?

As you can see, this is clearly an addition problem. Let's frame this question using a comparison bar model, as shown below:

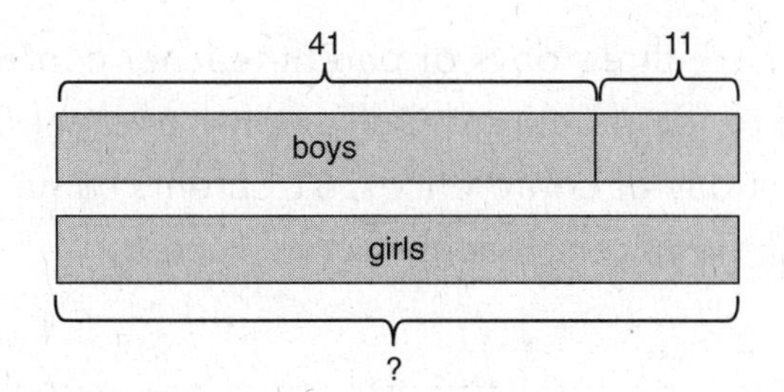

We are comparing the boys to the girls. We know the smaller quantity (boys). To find the bigger quantity (girls), we solve by adding: 41 + 11 = 52. This tells us that 52 girls signed up for the talent show.

Addition Exercises

1. What is 58 + 9?

- ○ A. 47
- ○ B. 48
- ○ C. 67
- ○ D. 57

2. What is 341 + 509?

- ○ A. 850
- ○ B. 840
- ○ C. 950
- ○ D. 805

3. Which statement is correct?

- ○ A. 413 + 613 = 816
- ○ B. 413 + 301 = 714
- ○ C. 701 + 161 = 956
- ○ D. 192 + 339 = 481

4. What is 543 + 399 + 42?

- ○ A. 984
- ○ B. 945
- ○ C. 1,014
- ○ D. 694

5. What is 41 + 32 + 99 + 14?

- ○ A. 192
- ○ B. 206
- ○ C. 184
- ○ D. 186

6. Last week there were three days of parent-teacher conferences. On Monday, 68 parents came to the meetings. On Tuesday, another 53 parents came in. On Friday, the last day of conferences, 61 parents came in. How many parents attended the parent-teacher conferences?

Answer: --

7. At High Mountain Elementary School, there are 127 third graders and 34 more fourth graders than third graders. How many fourth graders are at the school? Use the comparison bar model shown below to help you solve the problem.

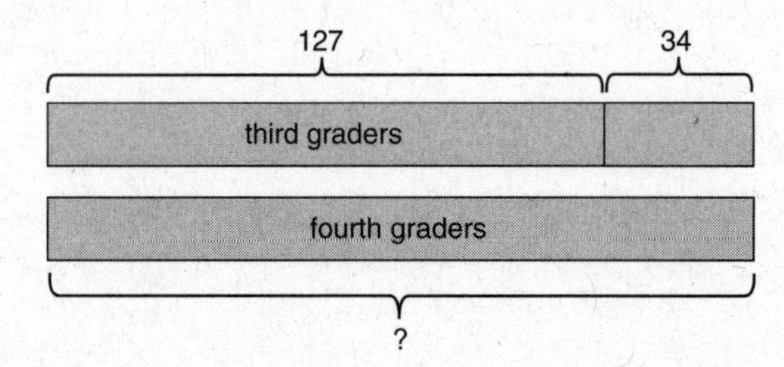

Explain your answer: --

----

8. At the supermarket, a box of cookies is $1.99. A gallon of milk is $3.39. How much will both items cost together?

- ○ A. $4.38
- ○ B. $5.38
- ○ C. $5.83
- ○ D. $5.46

9. Omar has 3 bird-feeders outside his house. One Saturday morning he decides to count the birds feeding. Bird-feeder #1 has 7 birds on it; bird-feeder # 2 has 11 birds on it; bird-feeder #3 has 4 birds on it. How many birds are feeding altogether?

- O A. 6
- O B. 23
- O C. 22
- O D. 27

10. Mrs. Sandel teaches 3 music classes. She has 24 students in the first class, 27 students in the second class, and 23 students in the third class, respectively. How many music students does Mrs. Sandel have in all?

- O A. 26
- O B. 71
- O C. 64
- O D. 74

Answers to Addition Exercises

1. **C.**

$$\begin{array}{r} {\scriptstyle 1} \\ 58 \\ +\ \ 9 \\ \hline 67 \end{array}$$

In this problem, 8 plus 9 is 17. Bring down the 7, carry the 1. 1 plus 5 is 6. So the answer is 67.

2. **A.**

$$\begin{array}{r} {\scriptstyle 1} \\ 341 \\ +\ 509 \\ \hline 850 \end{array}$$

In this problem, 9 plus 1 is 10. Bring down the 0, carry the 1. 1 plus 4 plus 0 is 5. 3 plus 5 is 8. So the answer is 850.

3. **B.**

$$\begin{array}{r} 413 \\ +\ 301 \\ \hline 714 \end{array}$$

4. **A.**

$$\begin{array}{r} {}^{1}\,{}^{1} \\ 543 \\ 399 \\ +\ \ 42 \\ \hline 984 \end{array}$$

5. **D.**

$$\begin{array}{r} {}^{1} \\ 41 \\ 32 \\ 99 \\ +\ 14 \\ \hline 186 \end{array}$$

6. 182 parents (68 + 53 + 61 = 182 parents)
7. 161 fourth grade students (127 + 34 = 161 fourth grade students)
8. **B.** $3.39 + $1.99 = $5.38.
9. **C.** 7 + 11 + 4 = 22 birds.
10. **D.** 24 + 27 + 23 = 74 students.

Subtraction

DEFINITIONS/VOCABULARY

Difference: the answer in a subtraction problem.

Subtraction: the process of taking one number away from another number.

Subtraction is the arithmetic process of separating or removing one value from another. Like addition, we use it every day using mental math. Subtraction is only slightly more difficult than addition. The main difference is that borrowing can be involved if the subtraction is not possible (for example, if you are trying to take a larger number from a smaller one, such as 3 – 7). We will look at that more difficult version shortly.

Subtraction is the opposite of addition, just as multiplication and division are opposites. If you have the problem 53 + 28 = 81, you can read it backward as 81 – 28 = 53. You will need paper and pencil becaues it requires borrowing. Again, like addition, you should ALWAYS rewrite the problem vertically before attempting to solve it. On exams you will often find it written horizontally (sideways, like a number model). It is important to rewrite it vertically, being careful to line up the place values properly.

$$\begin{array}{rcc} & 8 & 1 \\ - & 2 & 3 \\ \hline \end{array}$$

Just like addition, we line the numbers up on the right side, and then begin subtracting on the right-hand side. First, we will find 1 minus 3. The answer is NOT 2. Always read the problem going down, not up. We cannot take 3 from 1; the end result would be a negative number. Imagine if you had 1 piece of candy, and a friend wanted to take 3 pieces from you. It isn't possible. So to fix this, we must borrow. To borrow we take 1 from the digit immediately to the left. In this case, the next digit is the 8. So cross out the 8, make it a 7 (one less), and add 10 to the original 1 to the right.

$$\begin{array}{rcc} & 7 & 11 \\ & \cancel{8} & \cancel{1} \\ - & 2 & 3 \\ \hline \end{array}$$

That makes the new subtraction problem in the first column on the right 11 – 3. This would give us an answer of 8, so write 8 under the 3.

$$\begin{array}{rcc} & 7 & 11 \\ & \cancel{8} & \cancel{1} \\ - & 2 & 3 \\ \hline & & 8 \end{array}$$

Now we move to the left. 7 minus 2 is 5. So bring that 5 down.

$$\begin{array}{rcc} & 7 & 11 \\ & \cancel{8} & \cancel{1} \\ - & 2 & 3 \\ \hline & 5 & 8 \end{array}$$

There are no more digits to subtract to the left so our final answer is 58.

The borrowing process is by far the most common point of error. You must be very careful to make the neighboring digit one less when you borrow from it. Always show your work each step of the way to avoid costly mistakes.

Now that we have a firm understanding of the subtraction process, let's look at a word problem that uses subtraction. Remember, just knowing how to do the math is not enough. You need a deep understanding of how to apply these math processes through real life examples and word problems.

Example: At L. Housber Elementary School, there are 173 boys and 190 girls. How many more girls than boys are at the school?

A comparison bar model will help us solve this problem:

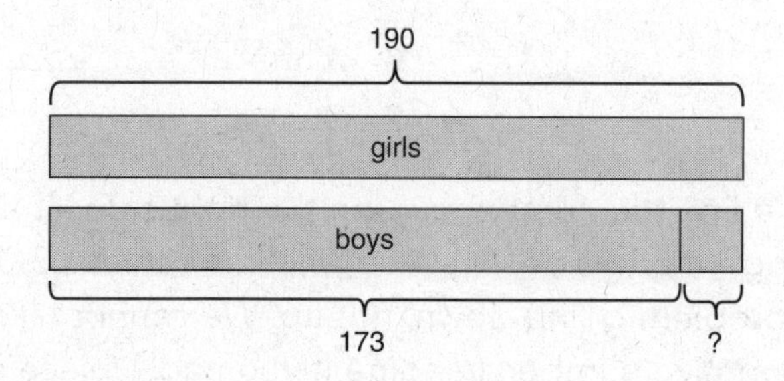

By looking at the comparison bar model, we can see that this is a subtraction problem to find the difference between the girls and the boys. To solve, we would compute 190 – 173 = 17. This tells us there are 17 more girls than boys at the L. Housber Elementary School.

Subtraction Exercises

1. What is 623 – 211?
- O A. 411
- O B. 412
- O C. 402
- O D. 311

2. What is 843 – 109?
- O A. 734
- O B. 746
- O C. 764
- O D. 736

3. What is 752 – 229?
- O A. 537
- O B. 573
- O C. 523
- O D. 533

4. What is 807 – 142?
- O A. 745
- O B. 565
- O C. 645
- O D. 665

5. What is 900 – 65?

- O A. 965
- O B. 865
- O C. 835
- O D. 825

6. At the supermarket a box of store-brand chocolate chip cookies is $1.99. A box of name-brand chocolate chip cookies is $3.79. How much more do the name-brand cookies cost than the store-brand cookies?

- O A. $1.08
- O B. $1.80
- O C. $2.20
- O D. $2.29

7. Justine needs to earn 30 points to get a free book. She already has 17 points. How many more points does she need to earn to get the free book?

- O A. 47
- O B. 23
- O C. 13
- O D. 27

8. Mr. Titus has a total of 142 students in his art classes. Each student needs to have his or her own paint brush. If Mr. Titus already has 113 brushes, how many more does he need to purchase?

- O A. 21
- O B. 31
- O C. 19
- O D. 29

9. Gale wants to expand his baseball card collection. He wants to have 450 cards total, but currently he has only 387. How many more cards does he need?

Explain your answer: --

----

10. Kristoff has 31 pencils and 13 pens in his desk. How many more pencils than pens does Kristoff have? Use the comparison bar model to assist you.

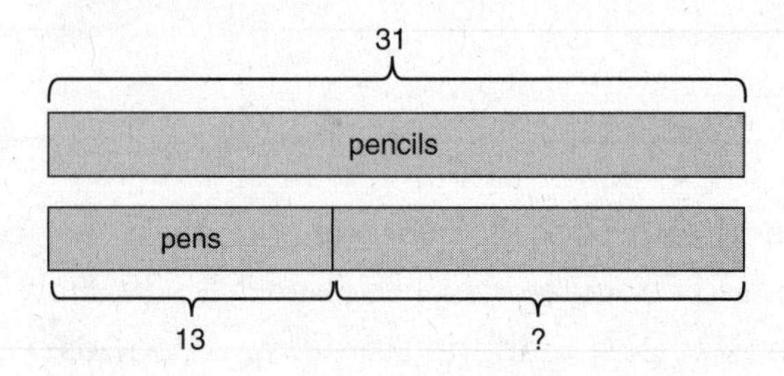

Explain your answer: --

----

11. Skylar has read up to page 212 in a novel that is 391 pages long. How many more pages will she need to read to finish the book?

○ A. 179
○ B. 279
○ C. 181
○ D. 81

Answers to Subtraction Exercises

1. **B.**

$$\begin{array}{r} \mathbf{6\ 2\ 3} \\ -\ \mathbf{2\ 1\ 1} \\ \hline \mathbf{4\ 1\ 2} \end{array}$$

This problem does not involve any borrowing because each time the number being subtracted (the bottom number) was smaller than the original (top) number.

2. **A.**

$$
\begin{array}{rrrr}
 & & 3 & 13 \\
 & 8 & \cancel{4} & \cancel{3} \\
- & 1 & 0 & 9 \\
\hline
 & 7 & 3 & 4
\end{array}
$$

In this problem we had to borrow from the 4 because we could not subtract 9 from 3.

3. **C.**

$$
\begin{array}{rrrr}
 & & 4 & 12 \\
 & 7 & \cancel{5} & \cancel{2} \\
- & 2 & 2 & 9 \\
\hline
 & 5 & 2 & 3
\end{array}
$$

In this problem we had to borrow from the 5 because we could not subtract 9 from 2.

4. **D.**

$$
\begin{array}{rrrr}
 & 7 & 10 & \\
 & \cancel{8} & \cancel{0} & 7 \\
- & 1 & 4 & 2 \\
\hline
 & 6 & 6 & 5
\end{array}
$$

In this problem we had to borrow from the 8 because we could not subtract 4 from 0.

5. **C.**

$$
\begin{array}{rrrr}
 & & 9 & \\
 & 8 & \cancel{10} & 10 \\
 & \cancel{9} & \cancel{0} & \cancel{0} \\
- & & 6 & 5 \\
\hline
 & 8 & 3 & 5
\end{array}
$$

This is the trickiest problem so far because we have to borrow twice. If we start on the right (as always) we see that we cannot subtract 5 from 0. Looking to the left is another 0 so we cannot borrow from it. Therefore we move again to the left and borrow from the 9. The 9 becomes an 8, and the middle 0 becomes a 10. Now that there is a 10 there we can cross it off and borrow from it to help the original 0. The 10 becomes a 9 and the original zero (all the way to the right) becomes a 10. Now we can subtract. $10 - 5 = 5$; $9 - 3 = 6$; $8 - 0 = 8$. The final answer is 835.

6. **B.** \$3.79 – \$1.99 = \$1.80; therefore, the name-brand chocolate chip cookies cost \$1.80 more than the store-brand cookies.
7. **C.** If Justine needs 30 points and she already has 17, we must subtract to find out how many more points she needs. 30 – 17 is 13. Don't forget to borrow because you cannot subtract 7 from 0!
8. **D.** If Mr. Titus needs 142 paint brushes and he already has 113, then we need to subtract to find out how many more paint brushes he needs to purchase. 142 – 113 = 29. Mr. Titus needs to purchase 29 paint brushes.
9. 63 cards. If Gale wants to expand his collection to 450 cards, but he only has 387 cards, we need to subtract to find out how many more cards he needs. 450 – 387 = 63 cards.
10. 18 pencils. If Kristoff has 31 pencils and 13 pens, we can see from the problem and the comparison bar model that subtraction will be used. To find the difference between the two numbers we solve: 31 – 13 = 18.
11. **A.** Skylar has read 212 pages of a 391 novel so we must subtract to find out how many more pages she must read. 391 – 212 = 179 pages.

Mixed Addition and Subtraction

Earlier we looked at some addition and subtraction strategies. We discussed these strategies separately to build a foundation for success with each concept. Sometimes, it will become necessary to use one process in order to solve what looks like the other. The example we gave in the addition section was 413 = 285 + ? We do *not* add 413 and 285 to solve this problem. Look closely at what it is asking. Remember, the equal sign means whatever is on the left-hand side must equal whatever is on the right-hand side. You may be used to seeing the sum, and the equals sign (=) on the right-hand side of the equation. Get out of that mindset. Remember that = means equal and does not have a specific set location for all math problems. Since 413 is by itself on the left, this tells us it is a sum. On the other side of the equals sign, we have 285, which is an addend, and there is a missing addend. How do we find the missing addend? We subtract. Think of it as a missing part. If there were 413 of something, and you took out 285, what would be left over? This is the missing addend. 413 – 285 = 128 (remember to write this vertically, as we learned above, because it involves borrowing twice). Therefore 128 is our missing addend.

Try a few practice problems.

Mixed Addition and Subtraction Exercises

1. 617 = 409 + ?

- O A. 1,026
- O B. 318
- O C. 218
- O D. 208

2. 290 = 70 + ?

- O A. 220
- O B. 210
- O C. 120
- O D. 360

3. 18 + 32 = ? + 14

- O A. 50
- O B. 26
- O C. 36
- O D. 40

4. 104 + ? = 511 – 200

- O A. 207
- O B. 307
- O C. 311
- O D. 104

5. 33 + 41 + 50 = 203 – ?

- O A. 170
- O B. 124
- O C. 69
- O D. 79

Answers to Mixed Addition and Subtraction Exercises

1. **D.** 617 is our sum; 409 is one of the addends, but we are missing the other addend. To solve, we must subtract 409 from 617 to find the missing part. 617 – 409 = 208. If you want to check to see if you are correct try it and see. Does 617 = 409 + 208? Yes, it does.

2. **A.** 290 is our sum; 70 is one of the addends, but we are missing the other addend. To solve, we must subtract 70 from 290. 290 – 70 = 220. Check by adding the two addends to see if the sum is 290. 70 + 220 is 290, so we are correct.

3. **C.** In this equation, we have a sum of 50 on the left-hand side of the equals sign (18 + 32). On the right-hand side we have a missing addend + 14, which must equal 50 (same as on the left-hand side). Therefore we can take 14 from 50 to find the missing addend. 50 – 14 = 36. Plug the 36 in and see if it equals 50. 18 + 32 is 50. 36 + 14 is 50. The total on the left-hand side equals the total on the right-hand side, so we are correct.

4. **A.** This equation is a bit more challenging, but it is not difficult. Remember, the total value on the left of the equals sign has to be the same as the total value on the right-hand side. On the right-hand side, we have 511 – 200. That equals 311. Now we must figure out a way to make the value on the left equal to 311. We have an addend there of 104, and a missing addend. We can take 104 from the sum (311) to find the other addend. 311 – 104 = 207. Plug it back in to check. 104 + 207 is 311; 511 – 200 is 311. Because the total value on the left-hand side is equal to the total value on the right-hand side, we are correct.

5. **D.** Let's start by adding up the total on the left-hand side of the equals sign. 33 + 41 + 50 is 124. Since we have the total, now we have to figure out how to make the right-hand side of the equals sign total 124 as well. We can subtract 124 from 203 to find the missing addend. 203 – 124 is 79. Check your work by plugging it into the equation. 33 + 41 + 50 is 124; 203 – 79 is 124. They are equal, so we know the answer is correct.

Multiplying by Multiples of 10

DEFINITIONS / VOCABULARY

Product: the result, or answer, when you multiply two or more numbers together.

Factors: the numbers being multiplied together.

In a multiplication equation such as $210 = 70 \times 3$ we have two important values.

In this example, 210 is the product, and 70 and 3 are the factors.

Earlier in this book we discussed properties of multiplication and different techniques to memorize the basic multiplication facts. Now that we have a firm understanding of those basic facts, it is time to extend them using multiples of 10. Third-grade students should be able to multiply a single digit factor by a multiple of 10, anywhere from 10 to 90 (for example, 3×30 and 5×70). This is a fairly straightforward process so let's first start with an example.

Example:

$$\begin{array}{r} 30 \\ \times \quad 5 \\ \hline \end{array}$$

In the problem, we are multiplying 30 by 5. If you look closely at this problem, you will see a basic multiplication fact: 3×5. In this type of fact extension, we can ignore the 0 in 30 and solve the basic fact of $3 \times 5 = 15$. Now that we have the basic fact solved, we can take the 0 and put it in the ones place to make a final answer of 150. Why does this work? Let's think of it this way. You're computing 30×5. When you multiply the 5 and the 3 you have to remember they each have a value. The 3 is in the tens position so you have 3 tens, right? Well if you multiply that times 5 you now have 15 tens. 15 tens is 150.

$$\begin{array}{r} 30 \\ \times \quad 5 \\ \hline 150 \end{array}$$

This method of fact extension only works when the numbers involved end with 0s. You can ignore the ending 0s, solve the basic fact, then put the 0s back into the final answer.

There is also a standard form of multiplication that will work with all multiplication problems. You can utilize this method to solve fact extensions (shown above) or standard multidigit problems. Let's take a look at another example using this method.

Example:

Step 1

Rewrite

Vertically

$$\begin{array}{r} 80 \\ \times \quad 3 \\ \hline \end{array}$$

In this problem we want to multiply 80 by 3. As in addition and subtraction, we always solve by starting on the right-hand side. The key to using this method is that the 3 has to multiply by every digit on the top in the correct order. First, we start with 3×0, which is 0. We place the 0 directly under the ones column because we just multiplied in the ones column.

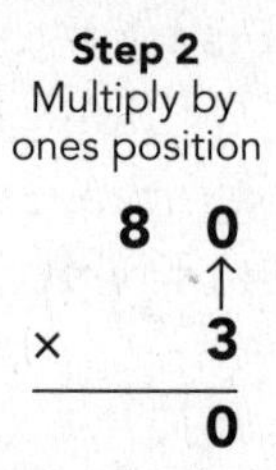

Next, the 3 must multiply by the 8. $3 \times 8 = 24$. As in addition, the 4 comes down into the tens position (since we just solved the tens position) and the 2 is carried over to the hundreds position.

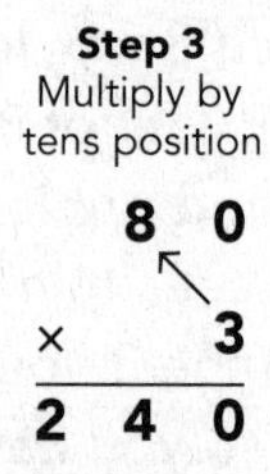

Finally, we come out with an answer of 240.

This method works just as well as the fact extension method we first discussed; however, the benefit is that it can be used as a foundational procedure for solving more complex multiplication problems in fourth grade. The fact extension method only works for problems where you have factors that include one digit 1-9 and all the rest are zeroes, whereas this method works for all multiplication problems, regardless of the number of digits.

Multiplying by Multiples of 10 Exercises

1. What is 7×90?
   - O A. 63
   - O B. 630
   - O C. 540
   - O D. 720

2. What is 80×5?
   - O A. 405
   - O B. 400
   - O C. 40
   - O D. 320

3. What is 9×10?

- ○ A. 900
- ○ B. 9
- ○ C. 90
- ○ D. 0

4. What is 70×7?

- ○ A. 560
- ○ B. 450
- ○ C. 400
- ○ D. 490

5. What is 5×60?

- ○ A. 300
- ○ B. 12
- ○ C. 3
- ○ D. 30

6. $3 \times 60 = ? \times 90$

- ○ A. 180
- ○ B. 20
- ○ C. 63
- ○ D. 2

7. The store clerk was stocking a shelf with cans of soup. The location on the shelf was designated for the soup. He found that the area on the shelf would hold 6 cans of soup from front to back and 20 cans of soup from left to right. How many cans of soup will the store clerk be able to fit on the shelf?

- ○ A. 26
- ○ B. 14
- ○ C. 620
- ○ D. 120

8. At an outdoor wedding reception, the staging company found that the area had room for 6 rows of chairs with 10 chairs in each row. How many guests will be able to sit in chairs?

Answer: --

9. Christopher has taken the time to arrange his marbles into an array. He has 4 rows of marbles with 20 marbles in each row. His friend Maria really likes the first 2 rows of marbles and wants to trade equally for them. How many marbles does she want to trade?

Answer: ____

10. Mr. Cucinotta owns a service station that specializes in replacing car tires. His shop is open 7 days per week. He estimates that he replaces 20 tires per day. If business continues at this pace then how many tires would he change in a week?

Explain your answer: ____

____

____

11. Mr. Martinez's house has a leaky roof and is in disrepair. He determines it is best to just replace the entire roof. The cost of new shingles and installation is calculated based on the area of his roof. The front of his roof measures 7 yards long by 10 yards wide. The back of his roof is the same. When Mr. Martinez calls to get price estimates, what should he say is the area of his roof?

Explain your answer: ____

____

____

Answers for Multiplying by Multiples of 10 Exercises

1. **B.** We can solve this using the fact extension method, which would be 7×9 (instead of 7×90) or 63, and then putting the 0 into the ones place. We come up with 630. We can also solve using standard multiplication:

STEP 1
Rewrite verically

$$\begin{array}{r} 9\;0 \\ \times \quad 7 \\ \hline \end{array}$$

STEP 2
Multiply by ones position

$$\begin{array}{r} 9\;0 \\ \uparrow \\ \times \quad 7 \\ \hline 0 \end{array}$$

STEP 3
Multiply by tens position

$$\begin{array}{r} 9\;0 \\ \nwarrow \\ \times \quad 7 \\ \hline 6\;3\;0 \end{array}$$

2. **B.** We can solve this using the fact extension method, which would be 8 × 5 (instead of 8 × 50) or 40, and then putting the 0 into the ones place. We come up with 400. We can also solve using standard multiplication:

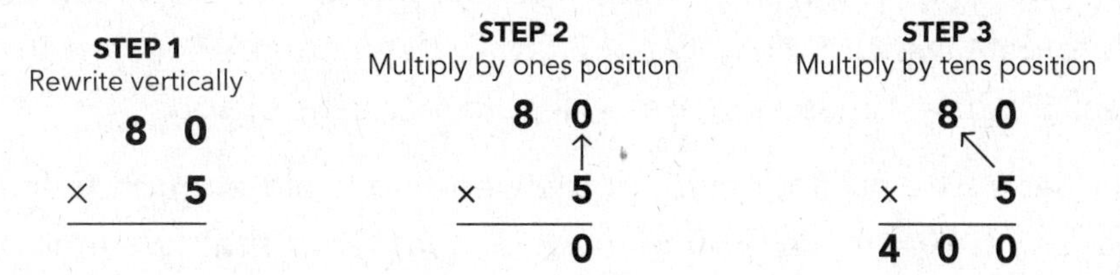

3. **C.** We can solve this using the fact extension method, which would be 9 × 1 (instead of 9 × 10) or 9, and then putting the 0 into the ones place. We come up with 90. We can also solve using standard multiplication:

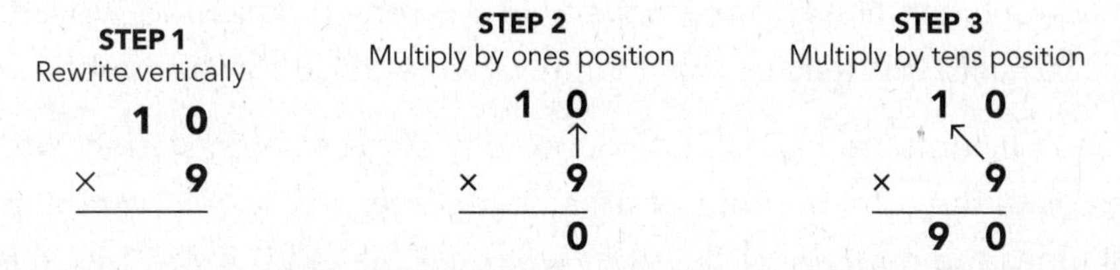

4. **D.** We can solve this using the fact extension method, which would be 7 × 7 (instead of 7 × 70) or 49, and then putting the 0 into the ones place. We come up with 490. We can also solve using standard multiplication:

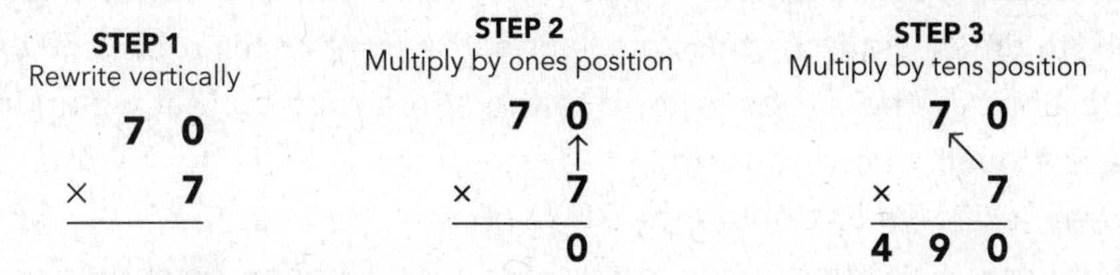

5. **A.** We can solve this using the fact extension method which would be 5 × 6 (instead of 5 × 60) or 30, and then putting the 0 into the ones place. We come up with 300. We can also solve using standard multiplication:

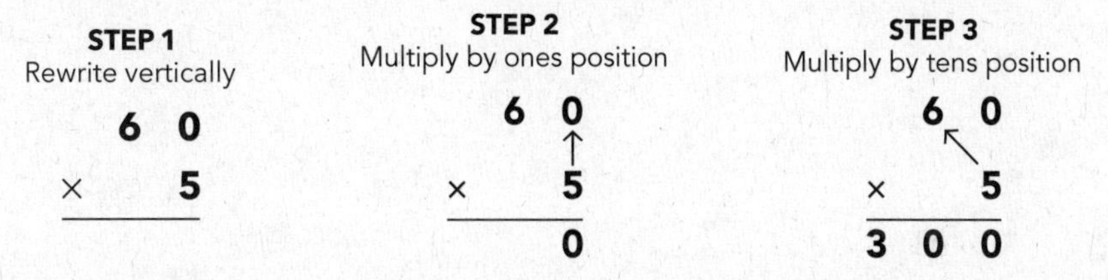

6. **D.** In this problem, we have an equation where the left-hand side must equal the right-hand side. The left-hand side equals 180 (3 × 60 is 180). So the question becomes, how do we make the right-hand side equal to 180 as well? The right-hand side gives us ? × 90. Remember, 180 is the product (total) so we have to find a factor to multiply by 90 (anthor factor) that equals

180. Thinking of your basic facts, what times 9 equals 18? It's 2. Since $2 \times 9 = 18$, 2×90 is 180, which is what we were looking for.

7. **D.** If the area on the shelf holds 6 cans from front to back, and 20 cans from left to right, we can multiply 6×20 to find the number of cans that will fill the area. $6 \times 20 = 120$, so the area will hold 120 cans of soup.

8. 60 guests. The outdoor area for the wedding is big enough to fit 6 rows of chairs with 10 in each. To find the total number of seats we multiply 6 by 10. $6 \times 10 = 60$. If there are 60 chairs, then 60 guests will be able to sit.

9. 40 marbles. Maria only wants to trade for the first two rows in Christopher's collection. The question states that each row has 20 marbles in it. If she only wants the first two rows, then we solve by multiplying 2 by 20. Since $2 \times 20 = 40$, this tells us that Maria wants to trade for 40 marbles.

10. If Mr. Cucinotta replaces 20 tires per day at his service station, we can calculate the total number of tires he replaces in a week by multiplying 20 times the number of days per week (7). Since $20 \times 7 = 140$, this tells us that if business stays on pace he will replace 140 tires per week.

11. This is a two-step word problem as we will see shortly. The front of Mr. Martinez's roof measures 7 yards $\times$ 10 yards. Area is calculated by multiplying length times width. By this calculation, the front of his roof is 70 yard^2. The question says he needs to replace the entire roof, so that would include the back as well. It also states that the size of the back is the same as the front. So we just have to double the 70 yard^2, which becomes 140 yard^2. When Mr. Martinez calls to get prices for his roof replacement, he should say that the area of his entire roof is 140 yard^2.

Number and Operations—Fractions

CHAPTER 5

DEFINITIONS/VOCABULARY

Denominator: the bottom value in a fraction. It refers to how many parts the whole is divided into.

Equivalent: equal in value.

Fraction: a number that names part of a whole or part of a group.

Numerator: the top value in a fraction. It refers to the number of parts you have.

Partition: to divide or split into smaller parts.

Unit Fraction: a fraction with a numerator of one.

Fractions—Parts of a Whole (Regions)

Fractions, like decimals, are frequently used to name part of a whole or part of a group. Let's start by taking a look at a region model to depict a fraction that represents part of a whole:

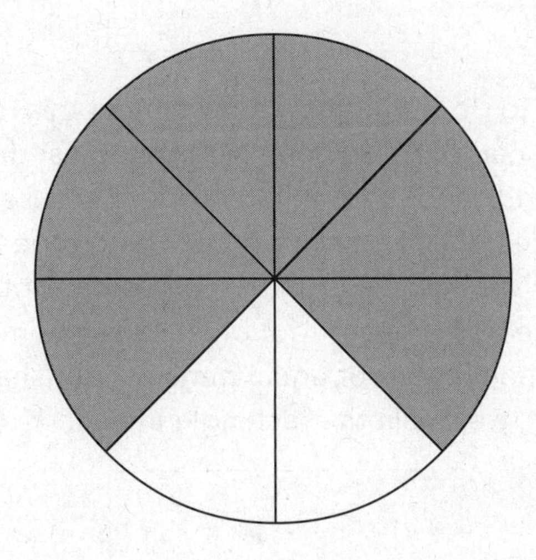

The region model above has been divided into 8 equal pieces. So in this example, the unit fraction would be $\frac{1}{8}$. Furthermore, we can see that 6 of those parts are shaded, while 2 are not. Before we discuss fractions pertaining to this circle, remember that the numerator (top number in the fraction) is the number of parts counted, and the denominator (bottom number in the fraction) tells the total number of equal parts. Looking at this circle we can come up with three distinct fractions:

1. The amount of the circle that is **unshaded** = $\frac{2}{8}$ (or 2 out of a total of the 8 equal parts)
2. The amount of the circle that is **shaded** = $\frac{6}{8}$ (or 6 out of a total of the 8 equal parts)
3. The entire circle as a fraction, based on the way it is divided into equal parts = $\frac{8}{8}$ (or 8 parts counted out of a total of 8 equal parts)

Let's take this one step further now. In the example given, we had a model, and we derived the fractions from the model. What if we look at this from another direction? Instead of making a fraction from a model, let's make a model from a fraction. In this example, let's shade in $\frac{1}{4}$ of the region.

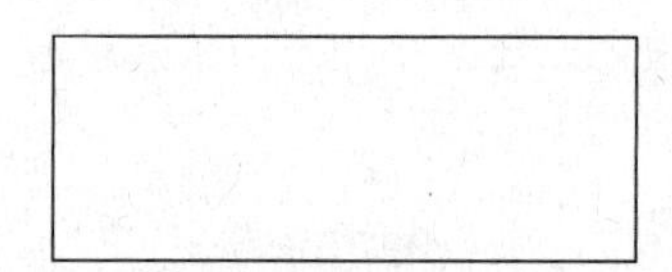

The first thing we need to do is figure out how to split this rectangle into fourths (since that's our denominator). Remember, we cannot just draw some lines to make 4 random parts. The parts need to be equal in size. Because we are dealing with fourths, which are simple fractions, one approach is to repeatedly divide the area in half. This will yield halves, fourths, eighths, and so on. Notice that each time you divide an area in half the denominator doubles. So each time you divide an area in half, you are doubling the number of equal parts in your shape. Let's see this approach in action. First, we'll split the rectangle in half.

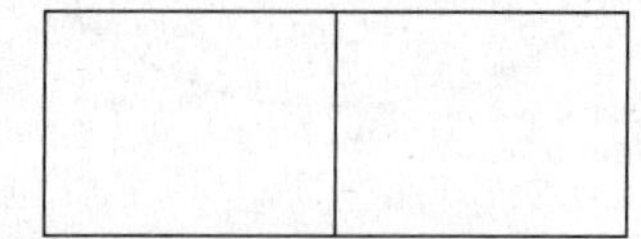

This is a good start, but we're not representing fourths yet. This rectangle currently represents halves. So let's split each of those halves into halves.

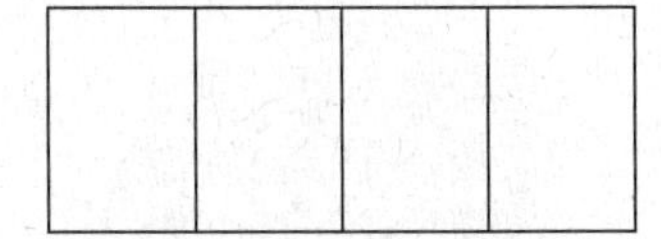

Now we have the rectangle broken into fourths (four equal parts). We have handled the denominator in our fraction $\frac{1}{4}$, so now it is time to focus on the numerator. The numerator is 1, and as you recall, the numerator is the represented portion from the original question. So we can wrap up this example by shading in one of the four equal parts:

$\frac{1}{4}$ shaded =

Fractions—Parts of a Whole (Regions) Exercises

1. In the figure below, which value would represent the fraction of the *shaded* region?

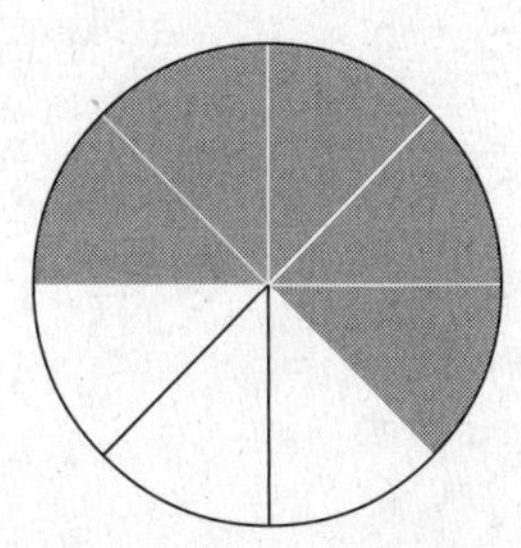

- A. $\frac{1}{3}$
- B. $\frac{3}{8}$
- C. $\frac{5}{8}$
- D. $\frac{8}{5}$

2. In the figure below, which value would represent the fraction of the *unshaded* region?

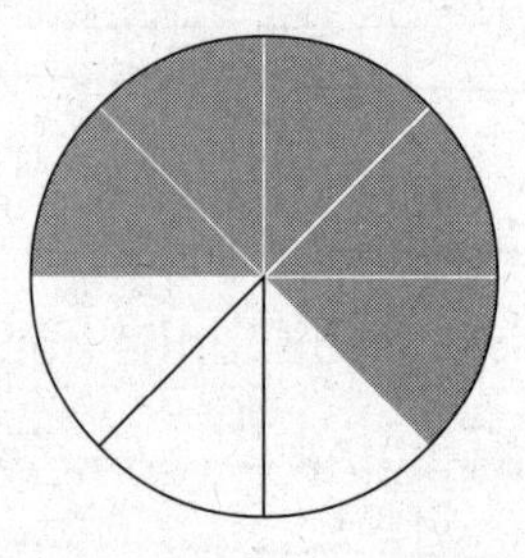

○ A. $\frac{5}{8}$

○ B. $\frac{8}{3}$

○ C. $\frac{3}{8}$

○ D. $\frac{1}{8}$

3. Victoria breaks a banana into 4 equal pieces. She eats one piece and gives the rest to her classmates. What fraction of the banana did Victoria eat?

○ A. $\frac{1}{4}$

○ B. $\frac{4}{4}$

○ C. $\frac{2}{4}$

○ D. $\frac{4}{1}$

4. Which circle shows a *shaded* region of $\frac{1}{4}$?

○ A.

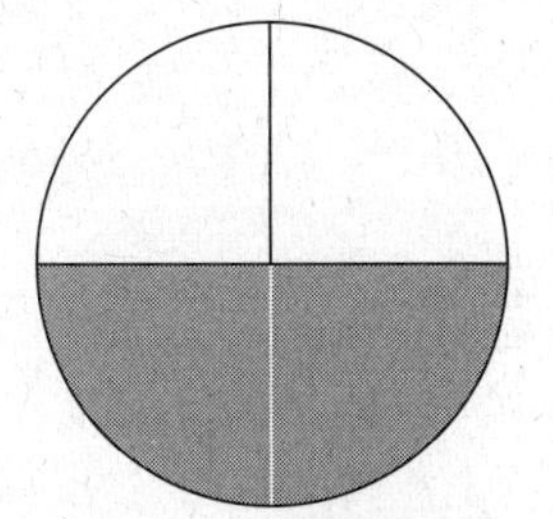

○ B.

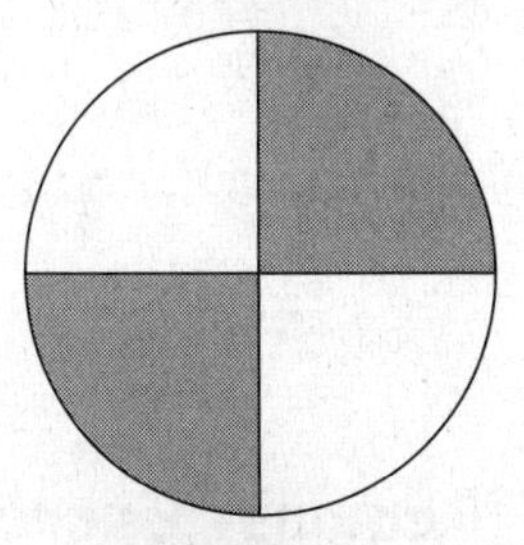

○ C.

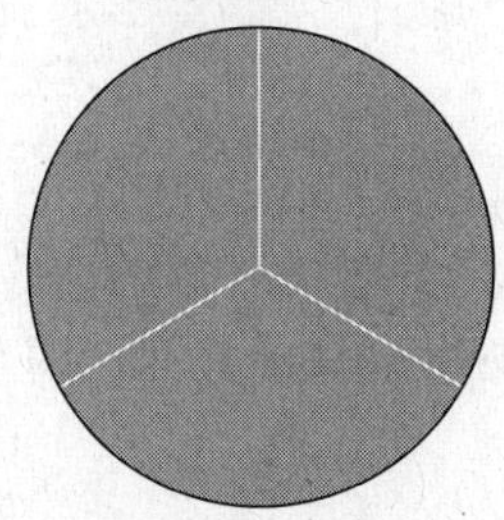

○ D.

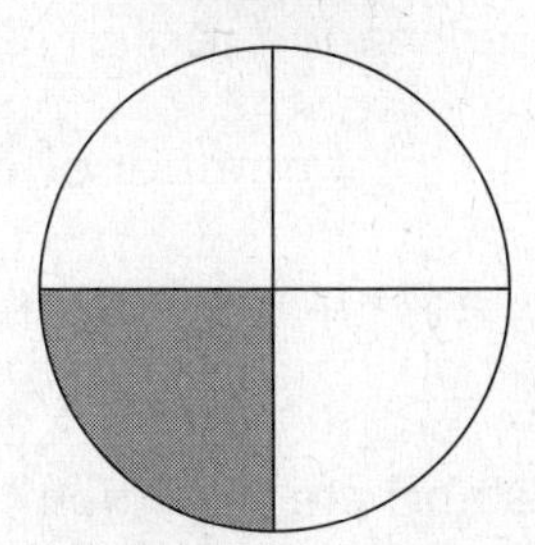

5. Shade in $\frac{3}{4}$ of the rectangle below.

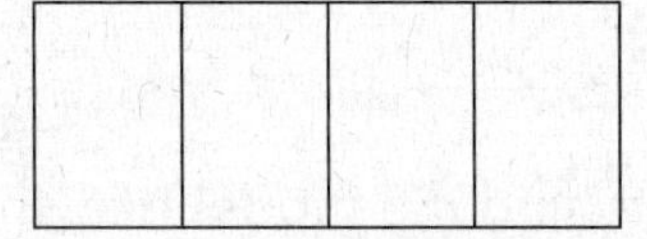

6. Divide the rectangle into the necessary number of equal parts and shade in to represent $\frac{2}{4}$ of it.

Explain your answer: --

Answers to Fractions—Parts of a Whole (Regions) Exercises

1. **C.** 5 out of 8 parts are *shaded*. Therefore, the *shaded* region represents $\frac{5}{8}$ of the circle. Remember, the denominator is the total number of equal parts (8) and the numerator is the number of equal parts we are referring to (the 5 that are *shaded*).
2. **C.** 3 out of 8 parts are *unshaded*. Therefore, the *unshaded* region represents $\frac{3}{8}$ of the circle. Remember, the denominator is the total number of equal parts (8) and the numerator is the number of equal parts we are referring to (the 3 that are *unshaded*).
3. **A.** Victoria broke the banana into 4 equal pieces and ate 1 of them. There are 4 equal pieces, so this is our denominator. Since we are referring to the number of pieces eaten, that will be our numerator (1). Therefore, Victoria ate $\frac{1}{4}$ of the banana.
4. **D.** The question asks us to find the model that shows a *shaded* region of $\frac{1}{4}$. The fraction means the total parts is 4 and 1 is shaded. Choice D is the only model that represents the fraction $\frac{1}{4}$.

5. This question asks us to shade in $\frac{3}{4}$ of a rectangle. Fortunately, the rectangle is already broken down into fourths. So this makes our job easy. We need to shade in three-fourths:

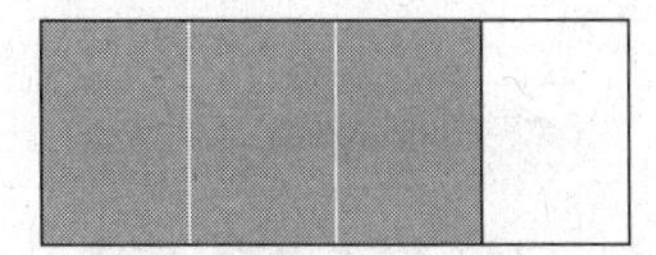

6. In this question, we are given a rectangle that contains only one equal part. The question tells us to shade in $\frac{2}{4}$ of it. We start by looking at the denominator to determine how many equal parts we should split it into. The denominator 4 tells us to break it into fourths, or four equal parts:

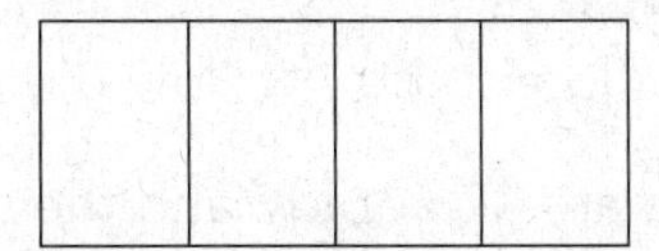

To get our final answer we need to look at the numerator (2) and shade in that number of equal parts:

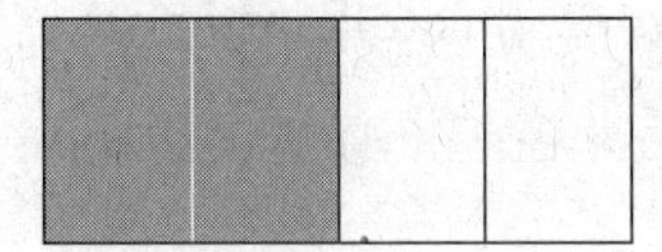

Fractions—Parts of a Group (Sets)

Just as fractions can represent part of a whole, they can also represent part of a group. For example, let's say we purchase a bag of apples at the supermarket. There are 12 apples in the bag (shown below):

What fraction of that group of apples is one apple? Well, we know that the numerator tells us how many we are counting, and the denominator tells us how many there are altogether. So one apple represents $\frac{1}{12}$ of the group of apples. Let's take it a step further:

Here we can see that part of our bag of apples is circled. Given the fact that this group of 12 apples is one whole, what would be the fraction that is circled? There are 3 apples circled, out of a total of 12 apples. So the fraction of apples circled would be $\frac{3}{12}$. The fraction could also be $\frac{1}{4}$. How did we come up with $\frac{1}{4}$? Look closely at what is given. There are 12 apples, and they are laid out in equal rows and equal columns (4 apples in each row; 3 apples in each column). Because of this, we can also represent the number of apples circled by thinking about the number of columns (since one whole column is circled). Since 1 column is circled out of a total of 4 **equal** columns, this tells us another fraction that shows the amount circled could be $\frac{1}{4}$ (or 1 out of 4). We will get into this concept in more detail later in the chapter when we move on to equivalent fractions. For now, let's work on creating some models.

As in the first section on parts of a whole, it's also important to be able to create a model to represent a given fraction. Let's walk through an example.

Example: Shade in $\frac{2}{4}$ of the rectangle.

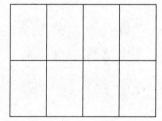

In this example, we have to first figure out what $\frac{2}{4}$ means. The rectangle is broken into 8 equal parts. The fraction $\frac{2}{4}$ tells us that we need to focus on 2 parts (the numerator) for every group of 4 parts (the denominator) we see.

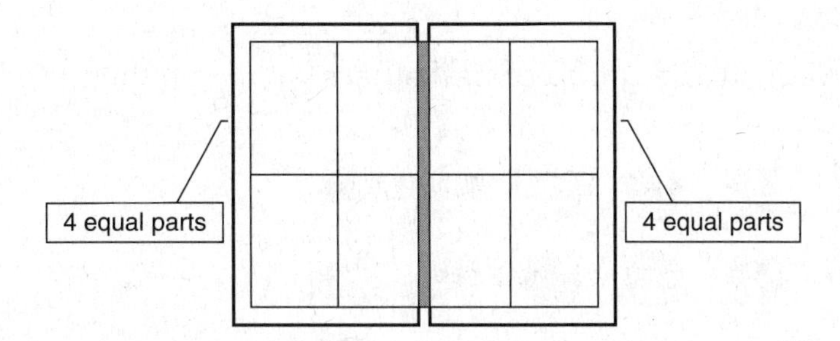

We can see from the diagram that there are 4 equal parts on the left side, and 4 equal parts on the right side. We can now focus on the numerator of the fraction $\frac{2}{4}$. It tells us to shade 2 out of every 4 so let's do that.

$\frac{2}{4}$ shaded = 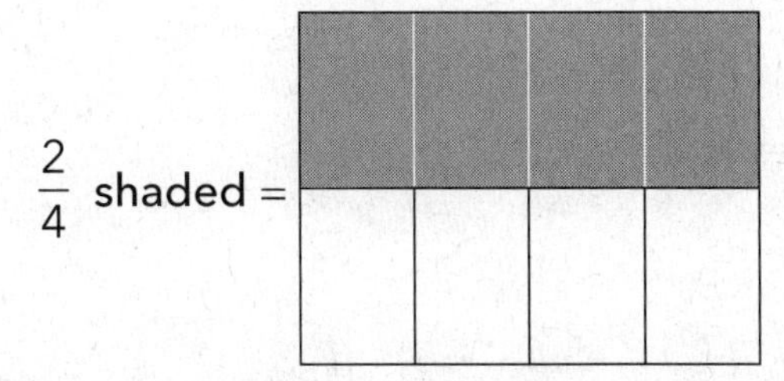

Now, let's try some practice problems, then we'll move on and take a look at fractions related to whole numbers.

Fractions—Parts of a Group (Sets) Exercises

1. Which fraction represents the circled portion of this group of strawberries?

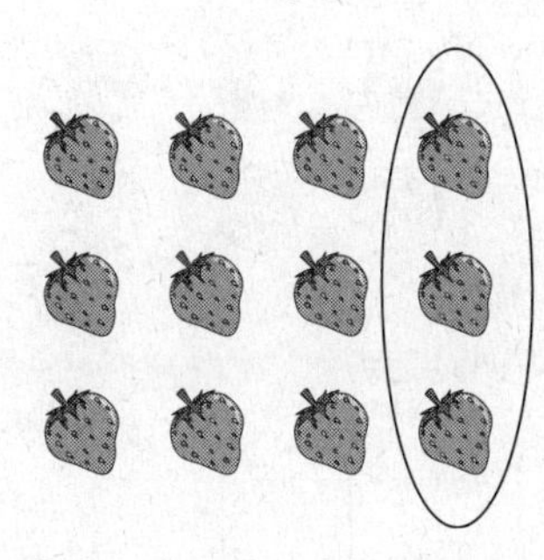

- ○ A. $\frac{2}{4}$
- ○ B. $\frac{1}{12}$
- ○ C. $\frac{3}{9}$
- ○ D. $\frac{1}{4}$

2. Which answer shows a group of basketballs with $\frac{2}{6}$ of them circled?

○ A.

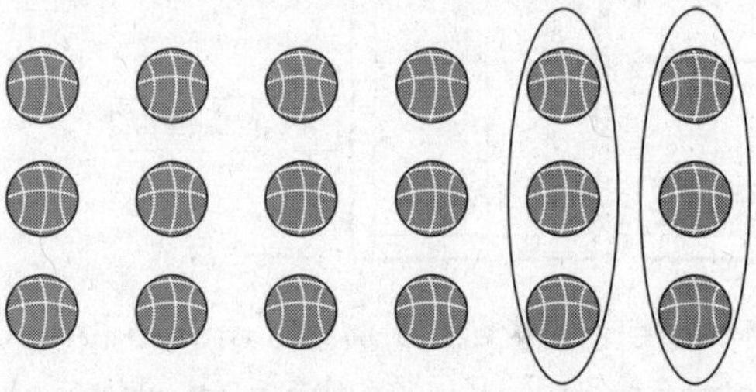

○ B.

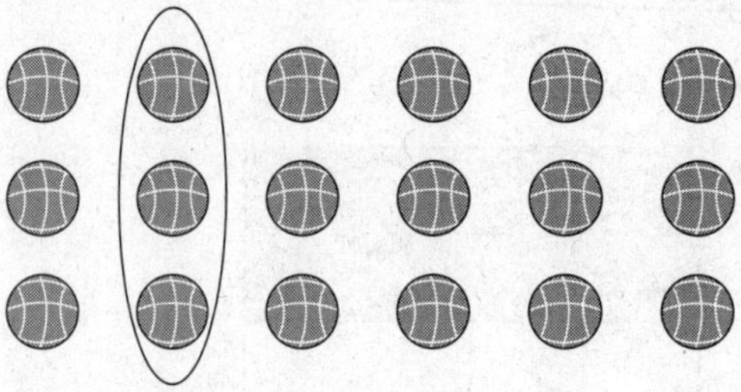

○ C.

○ D.

3. Shade in $\frac{2}{3}$ of the rectangle below.

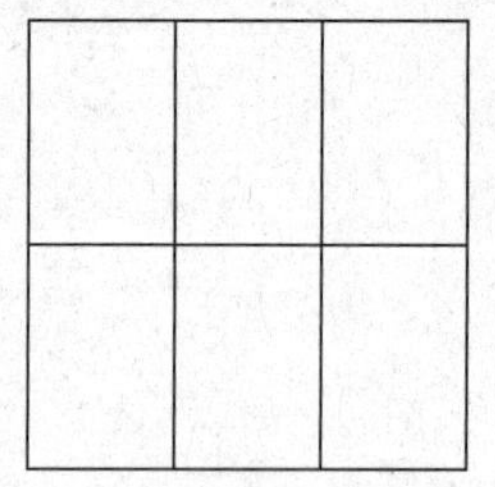

4. Shade in the rectangle below so that $\frac{1}{4}$ of it is *unshaded*.

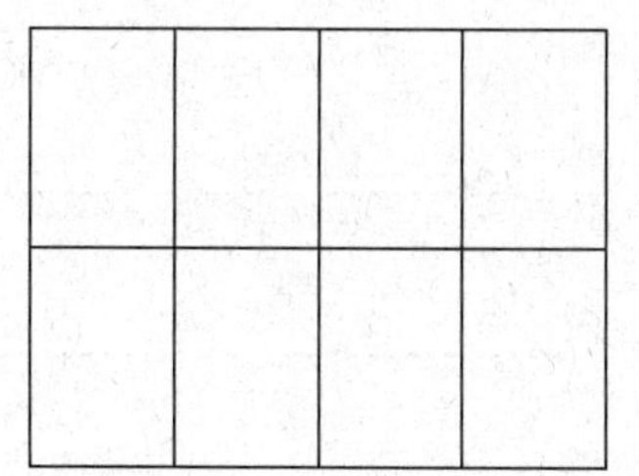

5. Divide the rectangle into the appropriate number of equal parts, then fill in the rectangle to show a *shaded* region of $\frac{3}{6}$.

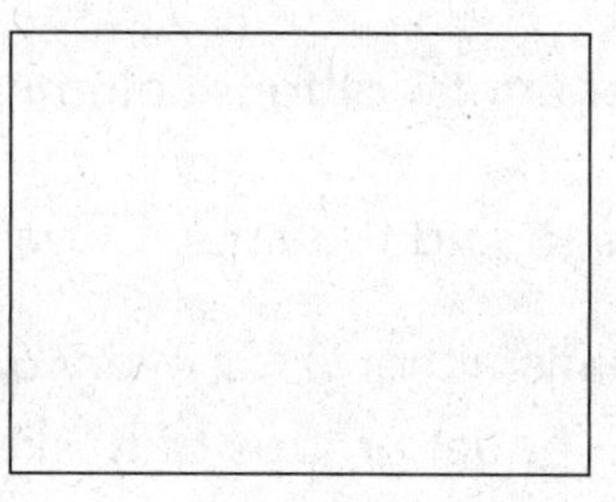

Explain your answer: --

----

6. The set model below shows a bag of lollipops that has been partitioned into equal groups. Each group is one whole.

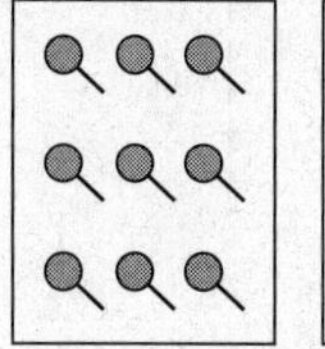
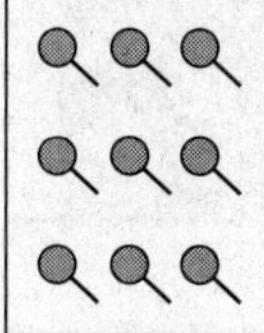
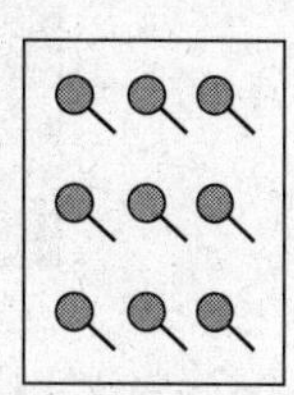

What fraction of a whole is each lollipop?

Answer: --

Explain your answer: ....

....

....

....

Answers to Fractions—Parts of a Group (Sets) Exercises

1. **D.** Looking at the strawberries, we can see there are a total of 12. The strawberries are split into 4 equal columns and 3 equal rows. The example shows one of those columns circled, so we will use that as our reference point. One column circled out of a total of four columns = $\frac{1}{4}$.

2. **A.** The question wants us to find the array of basketballs where $\frac{2}{6}$ of them are circled. Two out of six means either 2 for every 6, or two groups out of 6 equal groups. Choice A shows 6 equal groups of basketballs, with 2 of those groups being circled. This is the only correct answer.

3. In this question, we are asked to shade in $\frac{2}{3}$ of the rectangle. We can see by looking at the rectangle that there are 3 equal groups here represented as columns (based on our denominator). To solve, we can shade in 2 of those columns (based on our numerator).

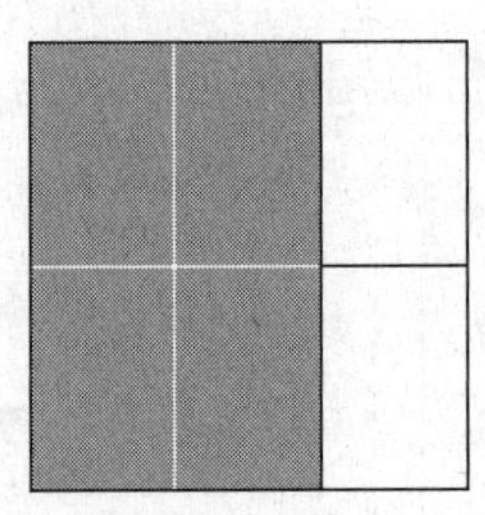

4. In this question, we are asked to model this rectangle so that $\frac{1}{4}$ of it is *unshaded*. We can see by looking at the rectangle that there are 4 equal groups here represented as columns (based on our denominator). Remember, we want 1 out of those 4 to be unshaded. That means we'll leave one column

blank, and shade the others. This will leave us with an unshaded region of $\frac{1}{4}$ and a shaded region of $\frac{3}{4}$.

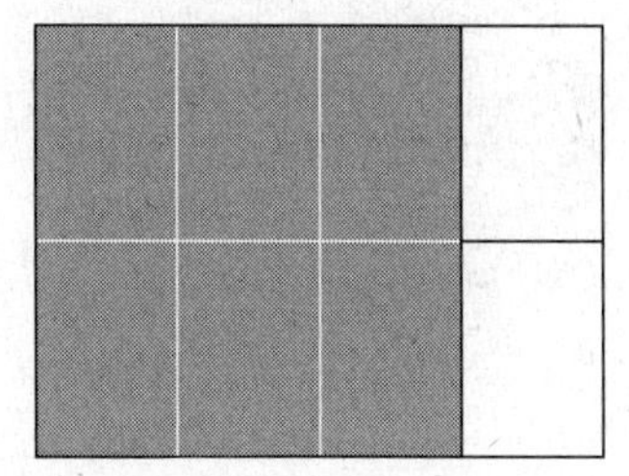

5. In this question, we need to divide a rectangle into sixths (our denominator) and shade in 3 of them. There are many answers for this question, but here is one example:

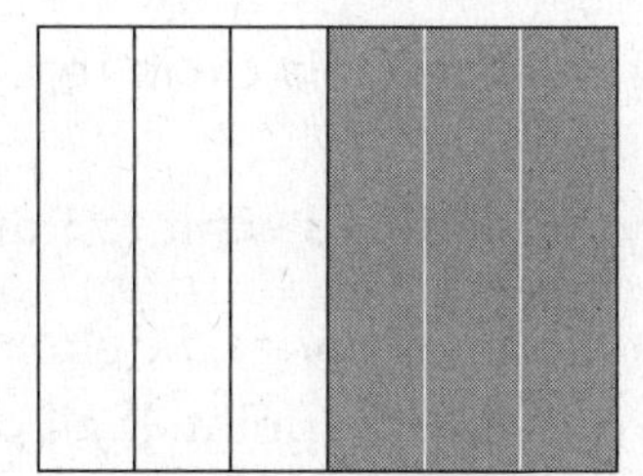

The key to answering this question correctly is that you must have 6 equal groups in your rectangle, and you must have 3 of those groups shaded.

6. $\frac{1}{9}$. Be careful on this one. The question is asking what fraction of a whole the lollipop is. It is asking in reference to a whole group, not a whole bag. You can see there are 27 lollipops there, but the question told us that a whole was one of the 3 groups–and each group has 9 lollipops. Therefore, one lollipop would be 1 out of 9, or $\frac{1}{9}$ of a whole.

Fractions—Related to Whole Numbers

Fractions are often misunderstood because they are seen as representing only a part of a whole or a part of a group. Let's take a more detailed look at how a fraction can be used to represent other whole numbers. Here is an example that shows one whole that is split into tenths. Note that in this example every part is shaded in.

Example: $1 = \frac{10}{10}$

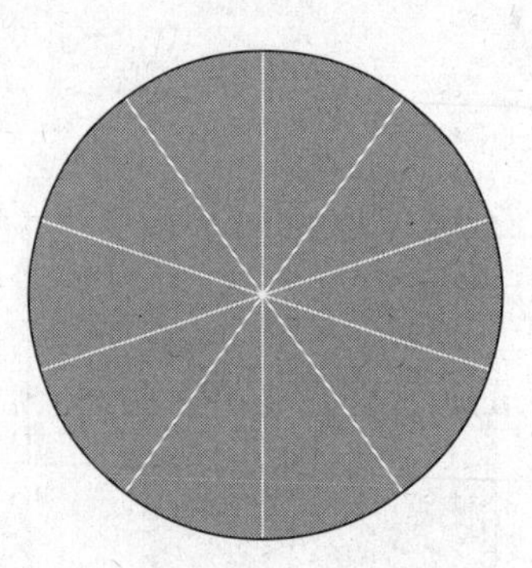

Based on this model, how do we know that 1 is equal to $\frac{10}{10}$? Well, look at the circle. It is divided into 10 equal parts and all 10 parts are shaded. We know that the numerator of the fraction is the counted parts, and the denominator is the total number of equal parts. Therefore, this circle represents $\frac{10}{10}$, also known as 1.

If you are given a fraction, say $\frac{6}{6}$, and you want to figure out if it is equal to 1, there is a simple solution. If the numerator and denominator are the same, it is equal to 1. The model above proves that point. If all parts of the shape/object are equal, and all are shaded, then that equals one whole shape/object.

In the previous example, we showed how a fraction can be equal to 1. Can a fraction be equal to other whole numbers, such as 2 or 3? Let's take a look at a model and find out.

Example: $3 = \frac{3}{1}$

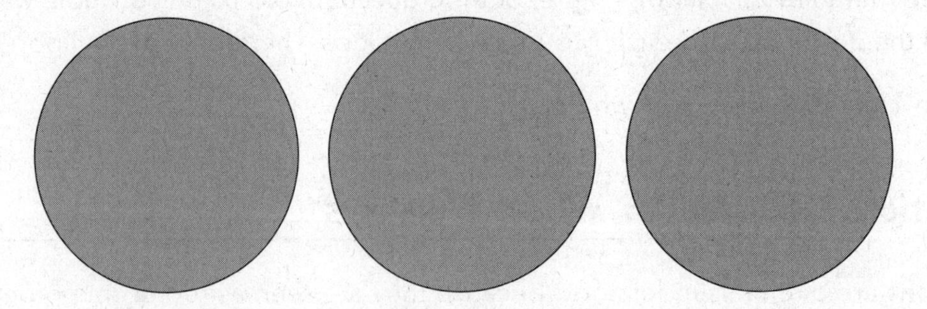

Notice in this example that we have 3 circles, all the same size. Each circle is completely shaded in, with only one equal part in each. Remember, denominators in fractions are the total number of equal parts in each shape. Since the circle only has one equal part, we know the denominator is 1. How many equal parts do we see between all 3 shapes? There are 3 equal parts. Each circle has one part, for a total of three. So our numerator is a 3. This makes our fraction $\frac{3}{1}$.

In the previous example, the shaded circles were represented by a fraction based upon the fact that each only had one shaded region (one equal part). Let's take this one step further and look at an example where we have multiple shapes split into multiple equal parts.

Example: $2=\frac{8}{4}$

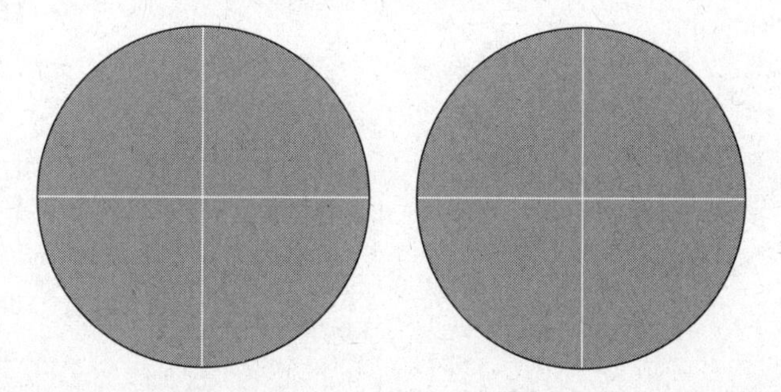

In this model we have two circles, both identical in size and both equally divided into fourths. All of the regions of these circles are shaded. So what number would represent these circles? You might be thinking "2", and yes, that is correct because there are two circles. However, a fraction also represents the circles. We noted before that the circles are identical and are divided into fourths. This tells us the denominator is a 4. Now, how many shaded fourths are there in all? There are 8 shaded fourths. So a fraction to represent the shaded region of these 2 circles would be $\frac{8}{4}$.

As in the previous example, this fraction may look odd because the numerator is larger than the denominator. In fact, because of this it is called an improper fraction. However, it is correct in its value and its representation. In fourth grade you will work to convert this improper fraction into proper form: a mixed number.

Fractions—Related to Whole Numbers Exercises

1. Which answer choice describes the fraction $\frac{6}{4}$?

○ A.

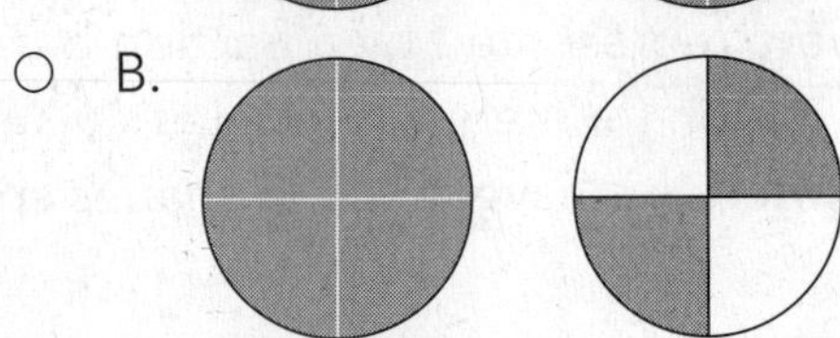

○ B.

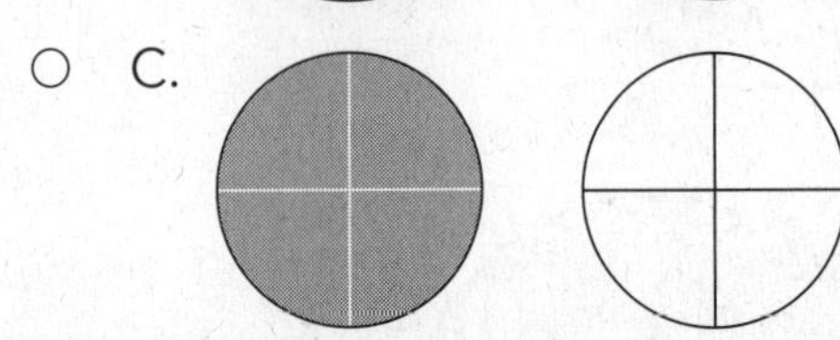

○ C.

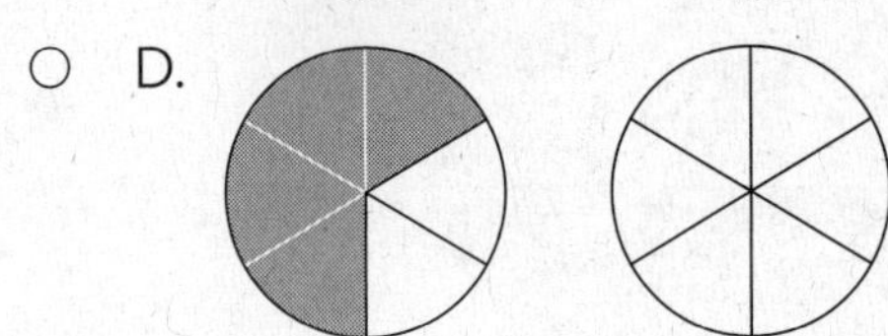

○ D.

2. Write a fraction for the shaded region.

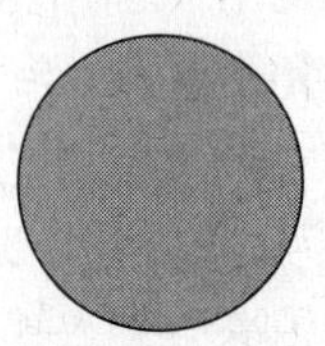

$$1 = \frac{?}{?}$$

Answer: ________________________________

3. Write a fraction for the parts that are shaded.

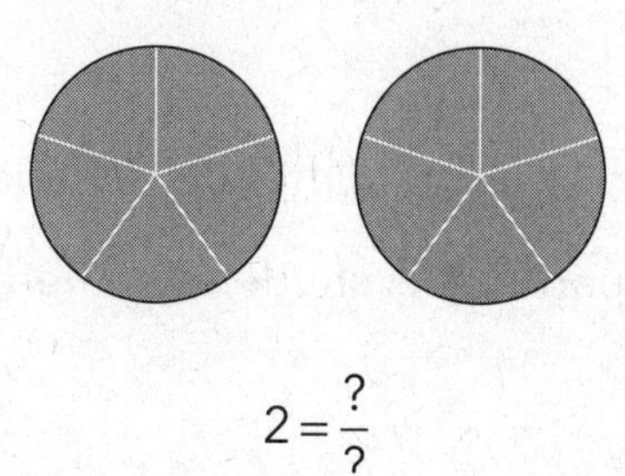

$$2 = \frac{?}{?}$$

Answer: ______________________

4. Write a fraction for the parts that are shaded.

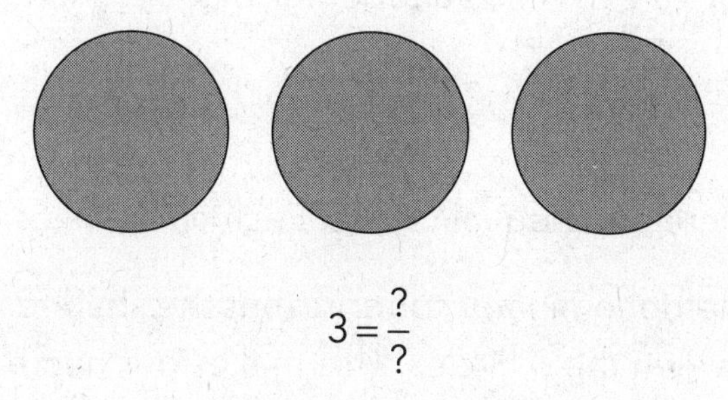

$$3 = \frac{?}{?}$$

Answer: ______________________

5. Shade the circles to represent the fraction $\frac{9}{6}$.

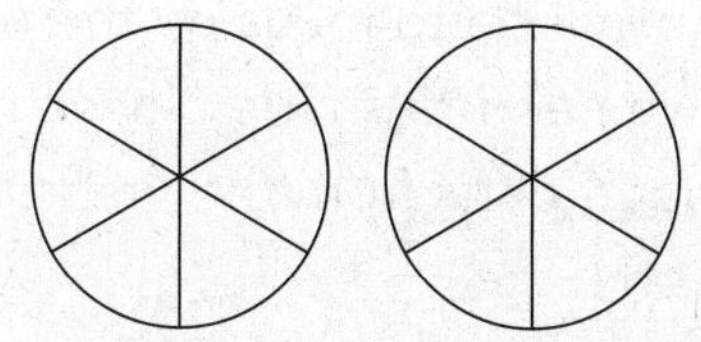

Explain your answer: ______________________

____

____

Answers to Fractions—Related to Whole Numbers Exercises

1. **B.** The question wants you to pick the model that represents $\frac{6}{4}$. The denominator 4 tells us the shape is split into 4 equal parts. The numerator tells us we are referring to 6 of those equal parts. One whole circle gives us 4 of

the parts, so we need 2 more parts from another circle. Choice B shows six-fourths being shaded in.

2. $\frac{1}{1}$. This question shows one circle completely shaded in. This means there is one part total, and one part that is shaded in. Therefore the shaded region is $\frac{1}{1}$, which is equal to 1.

3. $\frac{10}{5}$. This question shows two circles that are divided into fifths, and they are completely shaded in. Since they are divided into fifths, that is our denominator. There are 10 fifths altogether that are shaded in. So our fraction is $\frac{10}{5}$, which is equal to 2.

4. $\frac{3}{1}$. We can see there are 3 equal circles. Each circle has exactly one equal part in it. Therefore, the denominator to represent the shaded parts is 1. There are 3 equal parts between the 3 circles. This makes our numerator a 3. The fraction to represent the shaded parts is $\frac{3}{1}$.

5. In this question we are shading in $\frac{9}{6}$. The circles are already divided into sixths as you can tell. So our job will be to focus on the numerator 9. The numerator tells us how many parts we need to shade. So we need to count 9 parts and shade them in. We can get 6 parts from the first circle; then we'll need another 3 parts from the second one:

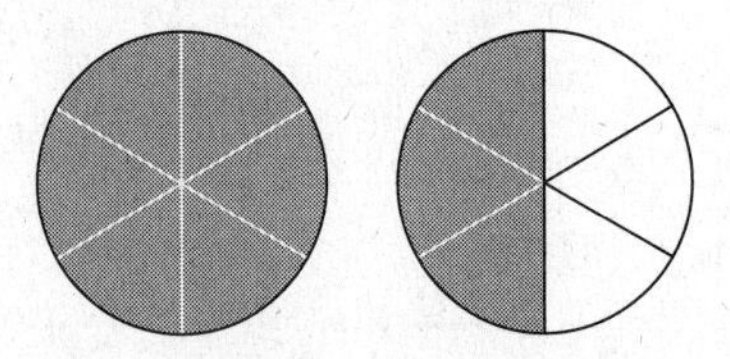

Fractions—Number Lines (Segments)

Another way fractions can be represented is in segments on a number line. A number line is a line that is divided into a certain number of equal parts. Each equal part is a segment and has a value that may be represented using a fraction. To determine the value of each fraction, you want to start at a whole number and count the equal spaces until you reach the next whole number. This will tell you the denominator (or total parts) for each of your fractions. Let's look at an example.

Example:

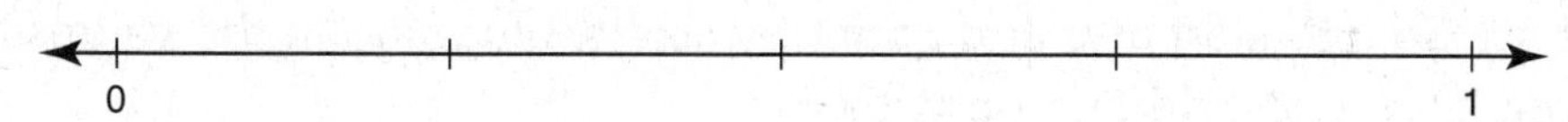

Looking at this number line, we can see that it starts at 0 and ends at 1. Since 1 is the next consecutive whole number after 0, we know that each of the marks on the number line has to be a fraction. To determine what fraction goes on each line, think back to the beginning of this chapter. Recall that a fraction is made up of a numerator and a denominator. The numerator is how many parts you are counting. The denominator is the total number of equal parts. How many equal parts are there from 0 to 1 on this number line? There are 4. That tells us the denominator is 4. Now we can fill in the number line with the values on each mark.

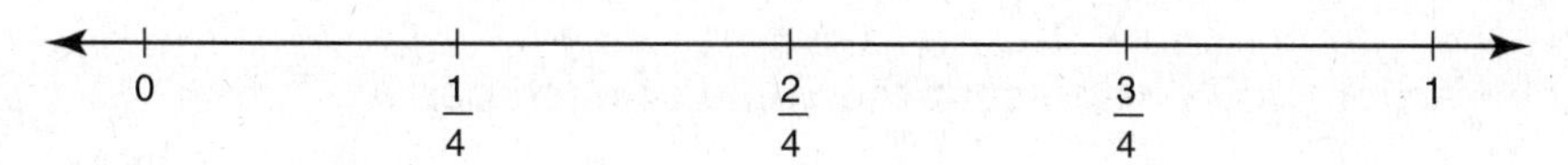

Each part on this number line is equal to $\frac{1}{4}$. So if we want to count up the value of each part as we go, we would start at 0 then count by fourths: $\frac{1}{4}, \frac{2}{4}, \frac{3}{4}$ and we would end at 1, which is the same value as $\frac{4}{4}$.

Additionally, sometimes with number lines we may find some of the marks are labeled with a value, while others are blank, so that you must determine and fill out the appropriate value.

Example:

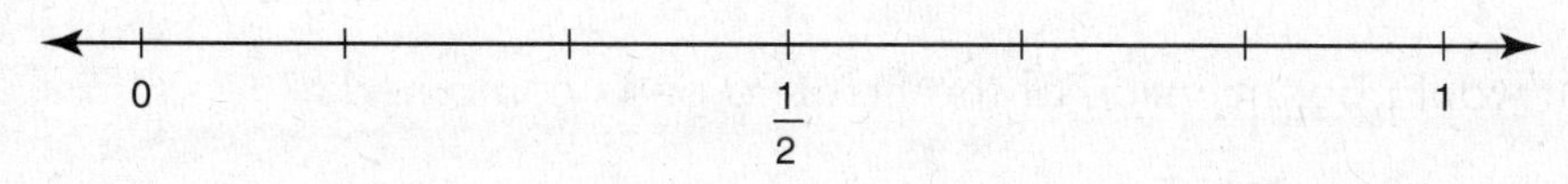

Looking at this number line, we can see what may be a little confusing. If you start at 0 and end at 1 you will find there are 6 equal parts. That tells us that the denominator in each fraction is 6. However, looking at what's given, we can see that there is a fraction on the number line that has a denominator of 2. There is a better explanation for this in the equivalent fractions section of this chapter. For now, remember, $\frac{1}{2}$ is half, or the middle, which you can clearly see. When we counted equal spaces between the two whole numbers 0 and 1, we found that there are six. So let's start at 0 and count by sixths, plugging in the values of the missing marks. You should end up with this:

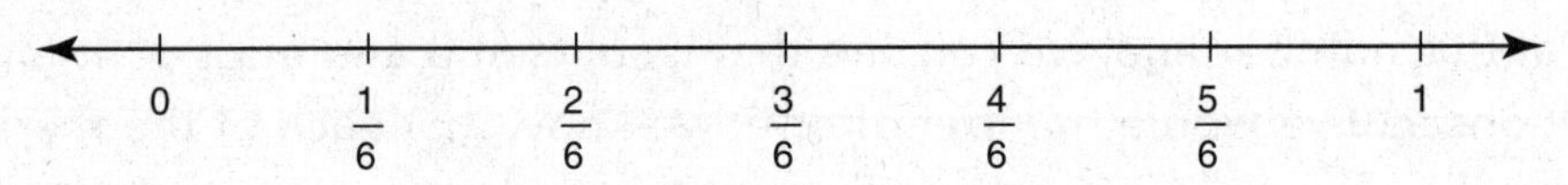

Fractions—Number Lines (Segments) Exercises

1. Look at the number line shown:

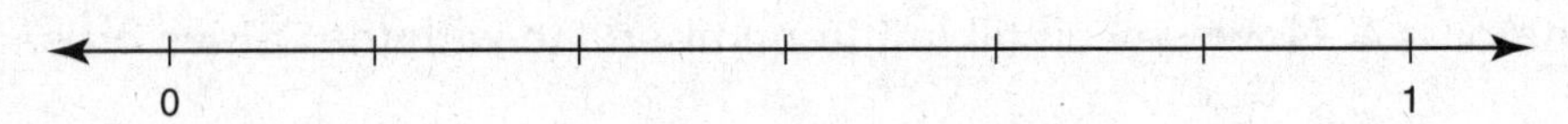

Which fraction would represent the value of each part?

- O A. $\frac{6}{1}$
- O B. $\frac{1}{1}$
- O C. $\frac{1}{3}$
- O D. $\frac{1}{6}$

2. Look at the number line shown:

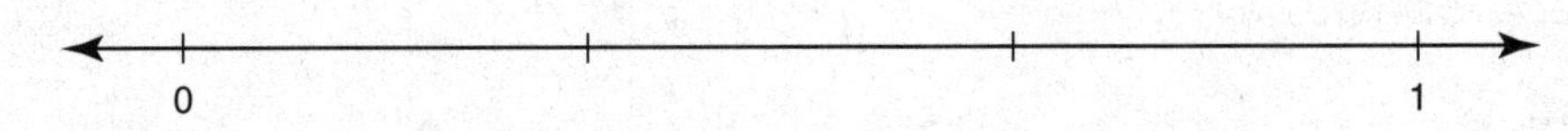

What would be the value of the first two parts combined?

- O A. $\frac{1}{3}$
- O B. $\frac{3}{2}$
- O C. $\frac{2}{3}$
- O D. $\frac{3}{3}$

3. Look at the number line shown:

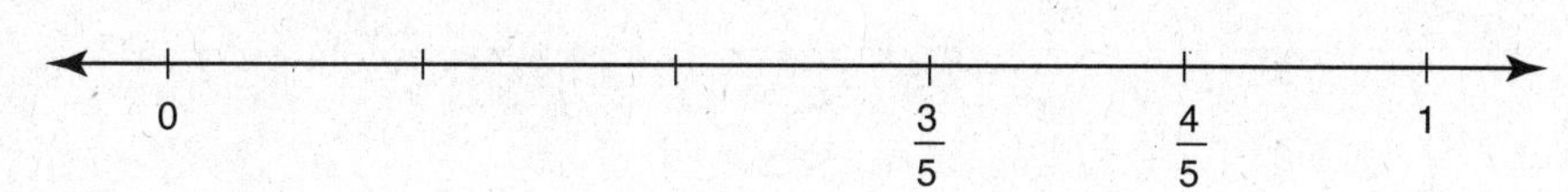

What fraction goes on the second blank line?

○ A. $\frac{1}{4}$

○ B. $\frac{1}{5}$

○ C. $\frac{2}{5}$

○ D. $\frac{5}{2}$

4. Kayla was asked to draw a number line that starts at 0, ends at 1, and is broken into sixths. This is what she drew. Is she correct?

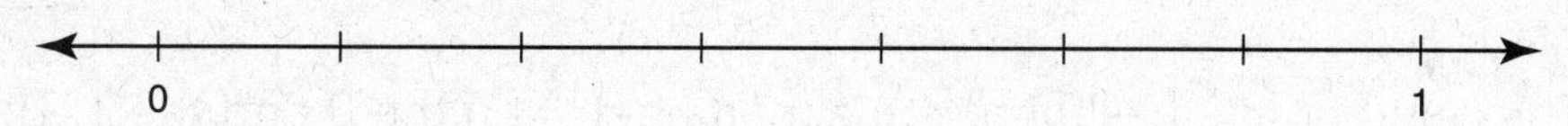

Explain your answer: --

----

5. Takim drew a number line that started at 0, ended at 1, and had 8 equal parts. What fraction would represent the total if we counted starting at the third mark all the way to the 1?

Explain your answer: --

----

6. What is the total for the shaded area?

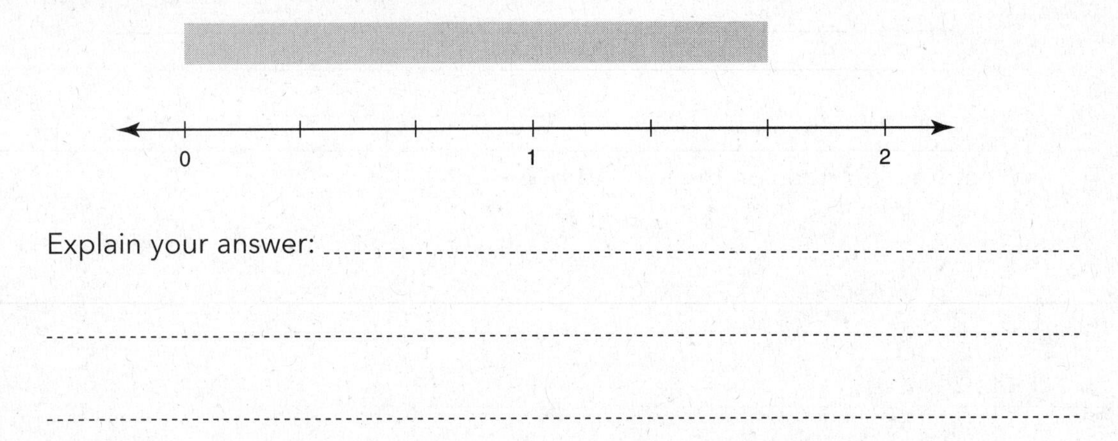

Explain your answer: --

Answers to Fractions—Number Lines (Segments) Exercises

1. **D.** If we start at the 0 and end at the 1 we can see that there are 6 equal parts. This tells us that each part is a sixth. Therefore, each part has a value of $\frac{1}{6}$.

2. **C.** This number line has three equal parts between the 0 and the 1. This tell us it is broken into thirds. The questions asks what the value of the first two parts is. If each part is a third, and there are 2 parts being counted, then their value is two-thirds or $\frac{2}{3}$.

3. **C.** This number line is broken into fifths. We can see that because there are 5 equal parts from 0 to 1. The third and fourth part are already labeled. Note that the question asks what fraction goes on the second blank line. Since this is broken into fifths, we just count by fifths two times and end up with two-fifths or $\frac{2}{5}$.

4. NO, Kayla's answer is not correct. The question stated that she should draw a number line from 0 to 1 that is broken into sixths. Instead of showing sixths, Kayla showed sevenths. We can see this by counting the number of equal parts from 0 to 1. Remember, count equal parts, not the marks on the number line.

5. $\frac{5}{8}$. On Takim's number line he is showing a total of 8 equal parts, going from 0 to 1. The question asks for the total if we start at the third mark and end at the 1. There are 8 equal parts from the 0 to the 1, so we know the

denominator is an 8. If we count up the equal parts from the third mark to the 1 we get 5. This tells us the total from the third mark to the 1 is five-eighths, or $\frac{5}{8}$.

6. $\frac{5}{3}$. This number line goes from 0 to 2. To figure out fractions we want to go to the next whole number and count the equal parts. If we go from 0 to 1 we can see there are three equal parts. So our denominator is 3. From the 1 to the 2 it is also 3 parts, so that is also broken into thirds. Now, looking at the shaded graph, we need to count how many parts on the number line are covered by the horizontal bar. We can see there are five-thirds or $\frac{5}{3}$.

Comparing Fractions (>, <)

Now that we have a better understanding of fractions, we can take the next step of comparing them to determine greater-than or less-than.

- Greater than is represented with the symbol >
- Less than is represented with symbol <

Remember, the open side of the symbol faces the greater fraction. Think of it as a little mouth that wants to gobble up the greater (larger in value) object. In third grade, when you compare fractions and want to know which is greater in value, or smaller in value, you will be faced with one of two situations, which we will discuss shortly. There is a strategy for solving each.

Fractions That Have Like Numerators

Let's look at $\frac{2}{3}$ and $\frac{2}{4}$. Both fractions have a 2 for a numerator. We know that the denominator tells the total number of parts in a whole. So let's think about this. If we make this into a real-life example with pizza, 2 slices from a pizza that's cut into 3 slices, is more than 2 slices from a pizza that's cut into 4 slices. This is because the slices from the 3-slice pizza are bigger in size than those from the 4-slice pizza.

$$\frac{2}{3} > \frac{2}{4}$$

Another way you can think about it is by looking closely at the fractions.

$\frac{2}{4}$ (2 out of 4) is the same as one half, correct? Look at the other fraction: $\frac{2}{3}$.

Is 2 more or less than half of 3? As you can see from the model, it's more.

Therefore, $\frac{2}{3} > \frac{2}{4}$.

Fractions That Have Like Denominators

Let's look at $\frac{1}{5}$ and $\frac{3}{5}$. Both fractions have a 5 for the denominator. Recall that the denominator is the total number of equal parts from the whole. This means there is an object split into 5 equal parts. These are the easiest fractions to compare because we just have to look at the numerator. 3 out of 5 is clearly more than 1 out of 5. Therefore, $\frac{1}{5} < \frac{3}{5}$. As proof, check out the model below:

$$\frac{1}{5} < \frac{3}{5}$$

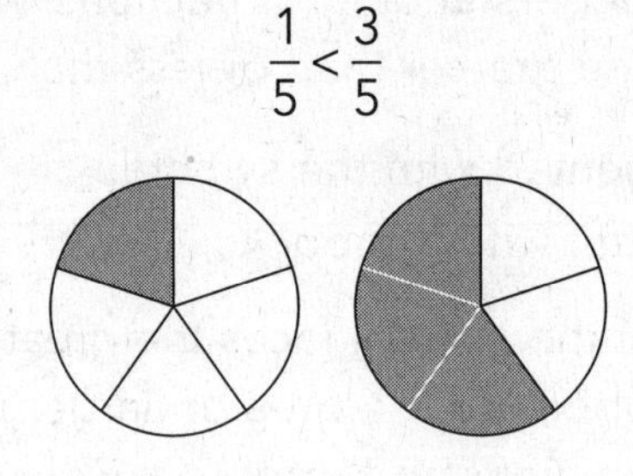

Comparing Fractions Exercises

1. Choose the correct symbol to compare the fractions:

$$\frac{4}{6} \_\_ \frac{2}{6}$$

O A. >
O B. <
O C. Neither of the above

2. Choose the correct symbol to compare the fractions:

$$\frac{3}{4} \_\_ \frac{3}{3}$$

- ○ A. >
- ○ B. <
- ○ C. Neither of the above

3. Choose the correct symbol to compare the fractions:

$$\frac{2}{2} \_\_ \frac{6}{6}$$

- ○ A. >
- ○ B. <
- ○ C. Neither of the above

4. Choose the correct symbol to compare the fractions:

$$\frac{4}{8} \_\_ \frac{4}{6}$$

- ○ A. >
- ○ B. <
- ○ C. Neither of the above

5. What fraction below is greater in value than the shaded area in this circle?

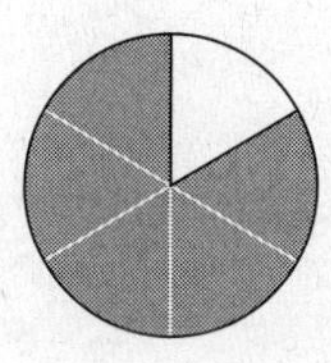

- ○ A. $\frac{1}{6}$
- ○ B. $\frac{2}{6}$
- ○ C. $\frac{3}{6}$
- ○ D. $\frac{6}{6}$

6. What fraction below is less in value than the shaded region?

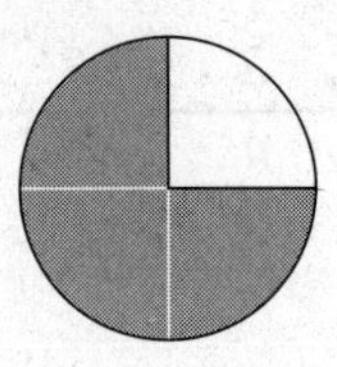

○ A. $\frac{2}{4}$

○ B. $\frac{3}{4}$

○ C. $\frac{4}{4}$

○ D. $\frac{5}{4}$

7. What fraction below is greater than the unshaded portion of this circle?

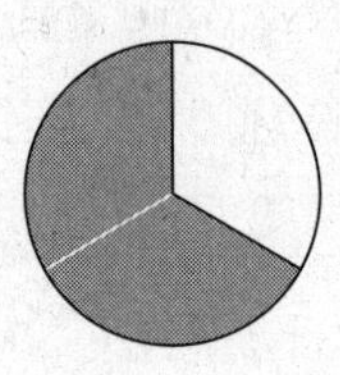

○ A. $\frac{3}{3}$

○ B. $\frac{1}{3}$

○ C. $\frac{1}{4}$

○ D. $\frac{1}{5}$

8. Which circle shows a shaded area less than $\frac{1}{2}$?

○ A.

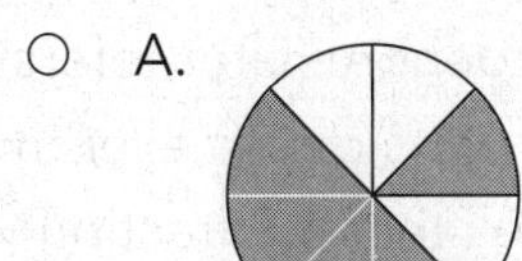

○ B.

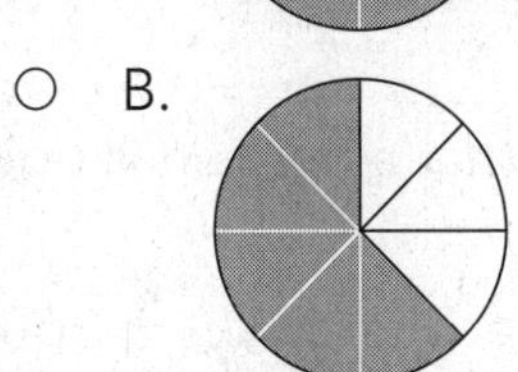

○ C.

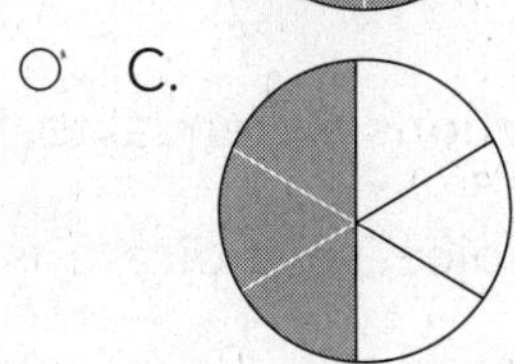

○ D.

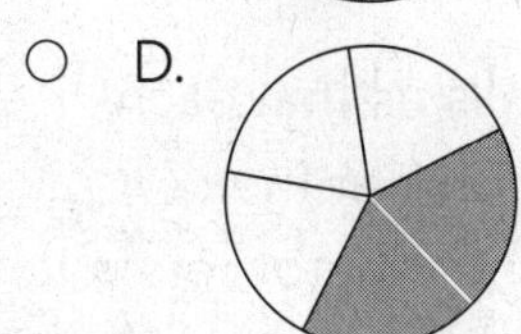

9. For breakfast, Bryan eats $\frac{3}{4}$ of a waffle. Maddie eats $\frac{1}{4}$ of a waffle. Which statement is true?

○ A. Bryan and Maddie ate the same amount of waffle.
○ B. Maddie ate more waffle than Bryan.
○ C. Bryan ate less waffle than Maddie.
○ D. Bryan ate more waffle than Maddie.

Answers to Comparing Fractions Exercises

1. **A.** In this question, we can see that the denominators are the same, which tells us they both represent a whole object with 6 equal pieces. Because of this, we need to focus on the numerators, 4 and 2. Since 4 is greater than 2, this tells us that $\frac{4}{6}$ is the greater fraction. So 4 pieces out of 6 is greater than 2 pieces out of 6.

2. **B.** In this question, we can see that the numerators are the same. This tells us that each fraction is referring to 3 parts of a whole or a group. Let's look at the denominators. We are comparing 4 and 3. Because the denominator refers to the total number of parts, as this number increases, the parts get smaller in size. Since the numerators are the same, we can easily tell that the fraction with the greater denominator is going to be the smaller one.

3. **C.** Let's look at the first fraction: $\frac{2}{2}$. We learned earlier that a fraction is equal to 1 when the numerator and denominator are the same. Therefore, $\frac{2}{2}$ equals 1. We can see the same thing with $\frac{6}{6}$. Because both fractions are equal to 1, they are actually equal in value. This is not an answer choice, so choice C is correct.

4. **B.** In this question, the numerators are the same. This tells us that each fraction is referring to 4 parts of a whole or group. Let's look at the denominators. We are comparing 8 and 6. Since the denominator refers to the total number of parts, as this number increases, the parts get smaller in size. Since the numerators are the same, we can easily tell that the fraction with the greater denominator is going to be the smaller one.

5. **D.** This circle is $\frac{5}{6}$ shaded in. The only fraction listed that is greater than $\frac{5}{6}$ is $\frac{6}{6}$. Think about it, $\frac{6}{6}$ would mean the entire circle is shaded in–6 out of 6 pieces. Therefore $\frac{6}{6}$ is obviously greater than $\frac{5}{6}$–or 5 out of 6 pieces.

6. **A.** $\frac{3}{4}$ of this circle is shaded in. The question asks us to pick a fraction that is less than that. $\frac{2}{4}$ is less than $\frac{3}{4}$, so this is our answer.

7. **A.** This circle has an area of $\frac{1}{3}$ that is unshaded. The only answer choice greater than $\frac{1}{3}$ is $\frac{3}{3}$.

8. **D.** In this problem, choices A, B, and C are all greater than or equal to $\frac{1}{2}$. You can clearly see this based on the shaded portions of each circle. Even though the shaded parts might not be next to each other in each choice, you can

easily visualize combining them to see that it is greater than or equal to $\frac{1}{2}$. If we look at choice D, we can see that it is clearly less than $\frac{1}{2}$; therefore, it is our only answer.

9. **D.** If Bryan ate $\frac{3}{4}$ of a waffle and Maddie ate $\frac{1}{4}$ of a waffle, we can compare these fractions using the strategy in this section. Both fractions have the same denominator, so the greater fraction is the one with the greater numerator. Since 3 is greater than 1, Bryan ate more waffle than Maddie.

Equivalent Fractions

Equivalent fractions are fractions that are equal in value even though they might look different. Even though the fractions look different, the overall values are the same. This may sound confusing, but it's actually quite simple. Let's shade in $\frac{1}{2}$ and $\frac{2}{4}$ to get a visual representation. In the figures below, we can see what each fraction looks like because we are representing it in a circle. Remember, the denominator (bottom number) is the total number of parts, while the numerator (top number) is the requested value (in this case, shaded parts).

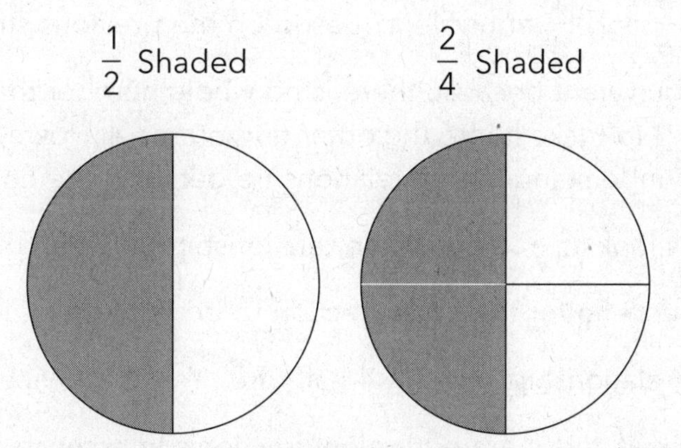

It's simple to see that these two fractions are equivalent because the shaded portion covers the exact same area in each. We shaded these circles in to mirror each other so that we can visually prove the values are in fact equivalent. The bigger question is: Why are they equivalent? Well, if we start with $\frac{1}{2}$, the denominator says that the object or group is divided into 2 parts. In the fraction $\frac{2}{4}$, the denominator says that the object or group is divided into 4 parts. So the total parts have doubled. Look at what happened to the numerator (shaded parts); it also doubled. Hence, doubling (×2), tripling (×3), and other multiplication are simple techniques to determine equivalency in simple fractions. When you apply the multiplication operation to both the numerator and the denominator, you are effectively increasing the number of parts at the same rate for both. This also works with division, again because the effect on the numerator and denominator is at the same rate.

In third grade, students are responsible for recognizing and generating simple equivalent fractions. For example, what if you wanted to figure out whether two fractions are equivalent but you do not have a visual representation? Let's use $\frac{1}{3}$ and $\frac{2}{6}$ as an example. The first thing you want to do is think about how you can make one of these fractions numerically turn into the other one using doubling, tripling, or other multiplication. In the previous visual the original fraction $\frac{1}{2}$ doubled. Can our fraction $\frac{1}{3}$ double to equal $\frac{2}{6}$? Absolutely. Doubling is the same process as multiplying by 2. If you multiply the numerator and the denominator by 2 you will end up with a new fraction, $\frac{2}{6}$. This tells us they are equivalent.

Another way to determine equivalent fractions is to relate the fractions to half. For example, are $\frac{3}{6}$ and $\frac{4}{8}$ equivalent? Based on the previous strategy, they would not seem to be equivalent because there is no whole number that would multiply by the numerator 3 to make it into the other numerator, 4. However, this alternate strategy relies on understanding the relationship between the numerator and the denominator. Let's look at $\frac{3}{6}$. What is the relationship between the 3 and the 6? 3 is half of 6, correct? So for the other fraction to be equivalent, it would also need to have the same relationship. In $\frac{4}{8}$, is 4 half of 8? Yes, it is. Since the relationship in each fraction is half, these are equivalent fractions. In essence, what you are doing here for each fraction is simplifying them to $\frac{1}{2}$, by dividing the numerator of each fraction by itself and its denominator.

This takes us to our next strategy. This is a more advanced strategy that is utilized in fourth grade, but it is a good alternative to have at your disposal. What we want to do is take each fraction down to its simplest form so we can more easily see if they are equivalent. (This strategy will also help you compare fractions for greater than and less than.) Let's simplify each fraction:

$\frac{3}{6} \rightarrow$ the numerator is 3, so divide the numerator and denominator by 3 to get $\frac{1}{2}$.

$\frac{4}{8} \rightarrow$ the numerator is 4, so divide the numerator and denominator by 4 to get $\frac{1}{2}$.

From this example, we can see that both fractions simplify down to $\frac{1}{2}$, so they are equivalent.

Let's go a step further. What if we want to *create* a fraction that is equivalent to another given fraction? In this case, we will either use multiplication (most common in third grade) or division to solve.

Example: Find a fraction that is equivalent to $\frac{2}{6}$.

This will be a two-step process. First, you must pick a whole number other than 0 and then multiply the denominator and the numerator by that number. This will create a new fraction that is equivalent to the old one. Let's randomly pick a whole number, say 3. *Note:* The lower the number you pick, the easier the multiplication process will likely be. So we multiply 2×3 to get 6 (the new numerator), and 6×3 to get 18 (the new denominator). We can now say that $\frac{2}{6}$ is equivalent to $\frac{6}{18}$. We can also use division to find an equivalent fraction. Is there any whole number that you can divide both 2 and 6 by? Sure, it's 2. $2 \div 2$ is 1 (the new numerator), and $6 \div 2$ is 3 (the new denominator. So using division, we've found that $\frac{2}{6}$ is also equivalent to $\frac{1}{3}$ (as we learned earlier).

Equivalent Fractions Exercises

1. What fraction below is equivalent to $\frac{3}{4}$?

○ A. $\frac{6}{8}$

○ B. $\frac{6}{5}$

○ C. $\frac{2}{3}$

○ D. $\frac{3}{8}$

2. What fraction below is equivalent to $\frac{5}{8}$?

- ○ A. $\frac{5}{24}$
- ○ B. $\frac{4}{16}$
- ○ C. $\frac{10}{4}$
- ○ D. $\frac{15}{24}$

3. What number goes in the blank space to make these two fractions equivalent?

$$\frac{5}{8} = \frac{?}{16}$$

- ○ A. 5
- ○ B. 11
- ○ C. 13
- ○ D. 10

4. Write a fraction that is equivalent to $\frac{3}{5}$.

Answer: --

5. List two fractions that are equivalent to $\frac{1}{2}$.

Explain your answer: --

----

6. List three fractions that are equivalent to the number 1.

Explain your answer: --

----

7. Two models of fractions are shown:

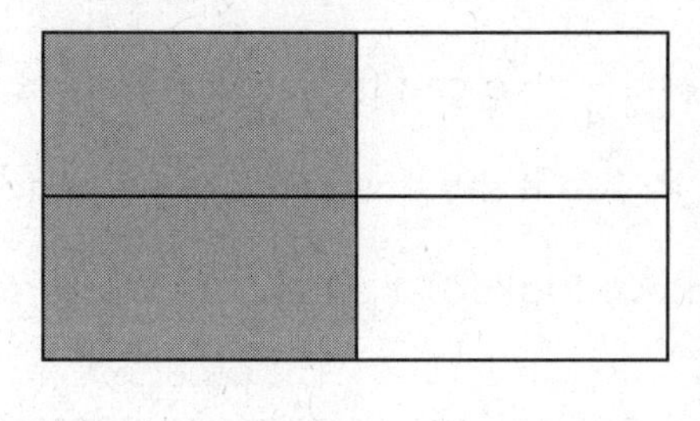

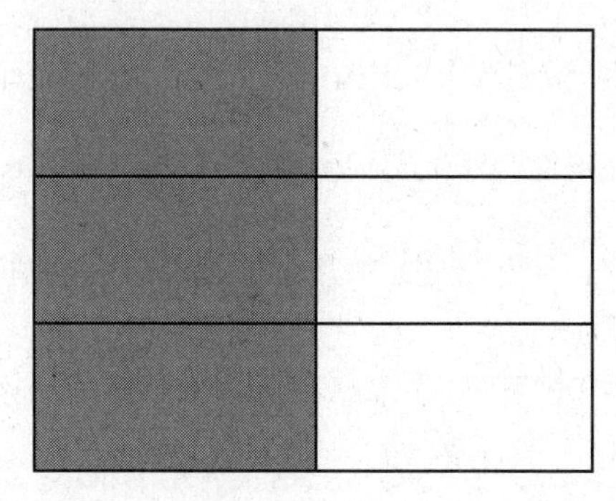

Write a number sentence using >, <, or = to compare the shaded regions of both rectangles.

Explain your answer: --

----

8. Two models of number lines are shown:

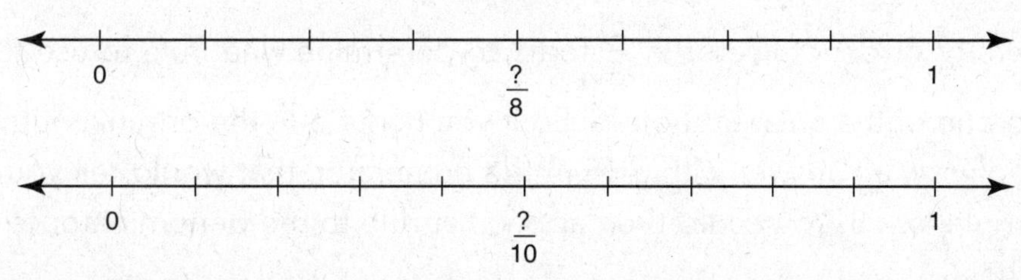

Determine the value of the two fractions shown and put them in a number sentence using >, <, or =.

Explain your answer: --

Answers to Equivalent Fractions Exercises

1. **A.** To find a fraction that is equivalent to $\frac{3}{4}$, we can try a few different techniques. First, we can try doubling to see if the answer is there. And in fact, it is. $\frac{3}{4}$ doubled is $\frac{6}{8}$. You could also have looked at either the numerator or denominator in $\frac{3}{4}$ to try to determine what rule to use to make it into one of the answer choices. So if you had a 3 in the original numerator, and you saw an answer with a 6 in the numerator, that would tell you the rule was double, or ×2. You could then apply that rule to the denominator to see if it matches.

2. **D.** To find a fraction that is equivalent to $\frac{5}{8}$, we can try a few different techniques. First, we can try doubling to see if the answer is there. $\frac{5}{8}$ doubled is $\frac{10}{16}$, which is not an answer choice. Next, you could try tripling. $\frac{5}{8}$ tripled is $\frac{15}{24}$. This is choice D. Alternately, you could also have looked at either the numerator or denominator in $\frac{5}{8}$ to try to determine what rule to use to make it into one of the answer choices. So if you had a 5 in the original numerator, and you saw an answer with a 15 in the numerator, that would tell you the rule was triple, or ×3. You could then apply that rule to the denominator to see if it matches.

3. **D.** If we look at the two denominators, we can see that the 8 can double to turn into 16. If we double the denominator, we must double the numerator. Therefore, the missing numerator is 10 (5 × 2).

$$\frac{5}{8}=\frac{10}{16}$$

4. $\frac{6}{10}, \frac{9}{15}, \frac{12}{20}, \ldots$ To find a fraction that is equivalent to $\frac{3}{5}$, we would pick a whole number and multiply the numerator and denominator by that number. Here are some examples: ×2 would give you $\frac{6}{10}$, ×3 would give you $\frac{9}{15}$, ×4 would give you $\frac{12}{20}$, and so on.

5. $\frac{2}{4}, \frac{3}{6}, \frac{4}{8}, \ldots$ There are an infinite number of possibilities for this question.

   Recall that to find an equivalent fraction you can pick a number and multiply the numerator and denominator by that number to get an equivalent fraction. In the answers given, I chose to do ×2, ×3, and ×4, respectively.

6. $\frac{3}{3}, \frac{7}{7}, \frac{9}{9}, \ldots$ There are an infinite number of possibilities for this question.

   Remember, for a fraction to be equivalent to 1, the numerator and the denominator must be the same. You can choose any number you want; just make sure you use the same number for both the numerator and the denominator.

7. $\frac{2}{4} = \frac{3}{6}$. In the first rectangle, there are 2 regions shaded out of 4. In the second rectangle, there are 3 regions shaded out of 6. We can see by the visual that the rectangles are the same size and the shaded regions are also the same size.

8. $\frac{4}{8} = \frac{5}{10}$. The top number line starts at 0 and ends at 1, and we can count 8 parts between them. Therefore, this number line is counting by eighths. If we count the marks we can see that the missing mark is the fourth one, so that fraction is $\frac{4}{8}$. On the second number line, we can see that it starts at 0, ends at 1, and has 10 parts in between. This tells us it is counting by tenths. Again, starting at the 0, we count five marks to reach the missing one. Therefore, that fraction is $\frac{5}{10}$. We can visually see that they are equal in value; therefore $\frac{4}{8} = \frac{5}{10}$.

Fractions—Real-Life Application

In the previous sections, we learned about all of the fundamental concepts pertaining to fractions in the third grade. Now, let's look at a few problem-solving questions as an extension. To answer these questions, you need to understand the concepts you have been taught and apply them via a word problem. In word problems, it is often up to you to select a strategy to solve the problem. In this way, we move into more concrete examples of fraction use in real-life situations. So in this section we will take more of a concrete approach to apply what we've learned. Here's an example of the type of question you may see in the practice set:

> **Mrs. Hodgens went to the supermarket earlier today. She purchased a container of blueberry muffins from the bakery department. The container had 4 muffins inside. Including Mrs. Hodgens, there are 6 people in the family. If they share the muffins equally, what fraction of muffin will each person get?**

It's always best to begin by modeling, so let's draw it out. You can pick any shape you want to represent the muffin, but for this example let's use a circle. Each circle represents one muffin from the container (4 muffins = 4 circles):

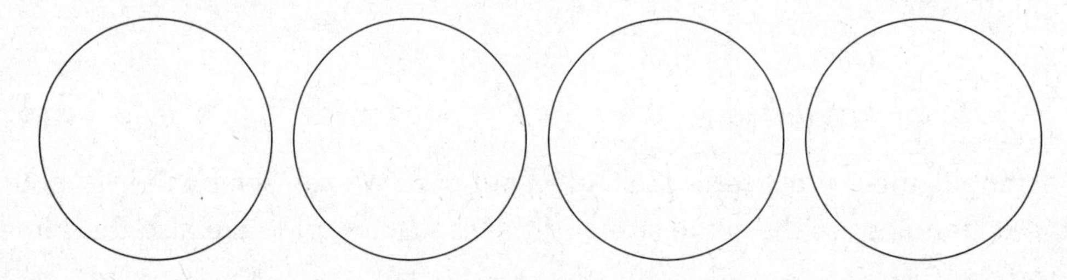

Next, we need to think about how to divide these circles (muffins). Each person is $\frac{1}{6}$ of the family, so we will want to divide each circle into sixths (6 equal parts):

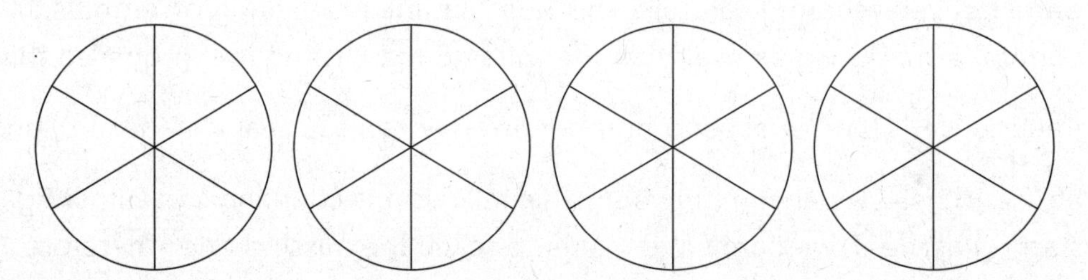

Now that the four muffins are divided into 6 equal parts (one for each family member) we can figure out the total for each family member by shading in $\frac{1}{6}$ of each circle (because each family member will get $\frac{1}{6}$ of each muffin).

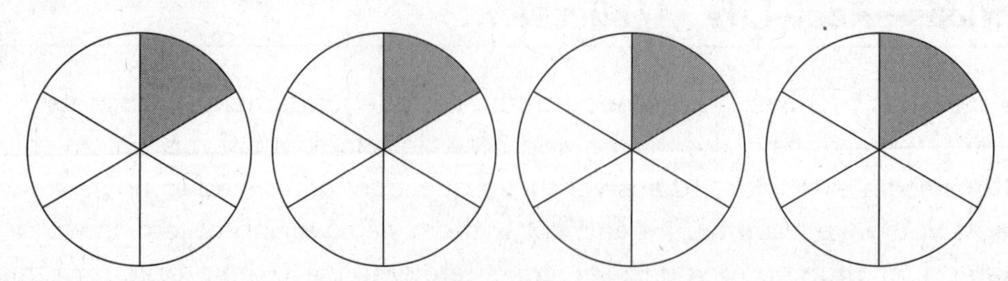

Finally, we can tabulate the grand total for each person. We can see that each person will be getting $\frac{1}{6}$ of each muffin. Since there are four muffins, that's a grand total of four-sixths, written as the fraction $\frac{4}{6}$.

Fractions—Real-Life Application Exercises

1. Isabella and her friend Jasmine decided to make peanut butter and jelly sandwiches for lunch. They also grabbed a bag of potato chips, and a glass of chocolate milk each. When Jasmine was finished making her sandwich, she cut it into 4 pieces and put it on her plate. Isabella cut her sandwich in half. If Jasmine ate 3 pieces of her sandwich, what fraction of the sandwich was remaining?

- ○ A. $\frac{3}{4}$
- ○ B. $\frac{1}{4}$
- ○ C. $\frac{1}{2}$
- ○ D. $\frac{4}{3}$

2. After a long day at school, Wendy arrived home. Before she had a chance to sit down, her mother told her she needed to go upstairs to clean her room. Wendy asked if she could first watch a half hour of TV. Her mother agreed. Afterward, Wendy reluctantly went upstairs and spent $\frac{1}{2}$ hour cleaning her room. Which fraction below is equivalent to $\frac{1}{2}$?

- ○ A. $\frac{3}{6}$
- ○ B. $\frac{3}{8}$
- ○ C. $\frac{2}{3}$
- ○ D. $\frac{2}{2}$

3. Trayvon brought in a bag of pencil erasers to share with his friends. Including himself, there were 6 children getting erasers. If the bag had 18 erasers in it, what fraction of the bag would each child get?

Workspace:

Explain your answer: ____

____

____

4. The Jacksons hosted a dinner party with a total of 12 people. For dessert, they brought out 2 pies. What fraction of pie will each person get?

Workspace:

Explain your answer: ____

____

____

5. Sandy and her friends shared a quesadilla for lunch. The quesadilla was split into 8 equal pieces. If they ate 7 pieces altogether what fraction of the quesadilla was left?

Workspace:

Explain your answer: ____

____

____

Answers to Fractions—Real-Life Application Exercises

1. **B.** This question purposefully has too much information in it. Weed through what is there to find out what you need. The question is only asking what fraction of Jasmine's sandwich is remaining. All of the other information is irrelevent. So let's figure out the denominator. She cut her sandwich into 4 equal parts, which makes our denominator 4. She ate 3 of those parts, so there is one remaining. Now we know our numerator, 1. The fraction of sandwich remaining, therefore, is $\frac{1}{4}$.

2. **A.** As in the previous question, this question has a lot of unnecessary information and wording. Sorting through the details, we can see there is a simple question: What fraction is equivalent to $\frac{1}{2}$? Recall that a fraction is equivalent to another fraction when we can multiply or divide the numerator and denominator by the same number to equal the other fraction. We cannot divide, because $\frac{1}{2}$ is already in its simplest form. So let's multiply: ×2 yields $\frac{2}{4}$, which is not an answer choice. ×3 yields $\frac{3}{6}$, which is choice A.

3. $\frac{1}{6}$ or $\frac{3}{18}$. Let's model this so we can understand the problem better.

   Remember, when you model fractions in an array, you want to make sure you are doing it in a way that leaves you with equal groups–that means equal numbers in each row and equal numbers in each column. The question states that 6 students would be getting erasers so we want to make sure we have either 6 equal rows or 6 equal columns.

Now that we have an array representing the 18 pencil erasers, it is that much easier for us to see what fraction each of the 6 children will get. This array is laid out with 6 equal columns, so each column can be equal to what one child receives. So let's circle a column. That will show what fraction of the erasers each child gets:

Finally, we can see that each child will get $\frac{1}{6}$ of the package of erasers (1 column out of 6). It could also be written as $\frac{3}{18}$ (3 erasers circled out of 18 erasers total).

4. $\frac{2}{12}$. If the Jackson family is sharing 2 pies with 12 people you could begin by modeling the pies and breaking up each into 12 equal pieces. Then each of the 12 people would get one slice from each pie.

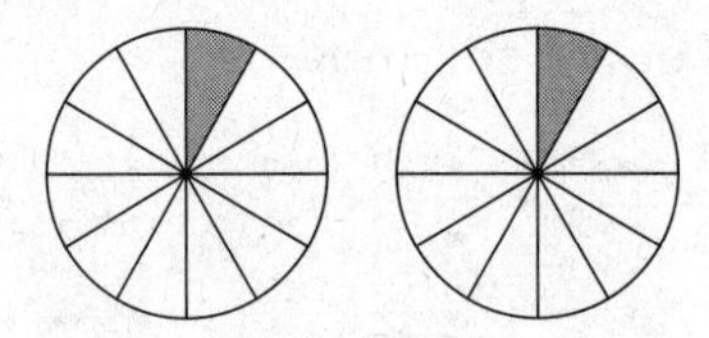

Since each pie is broken into 12 equal pieces we can count by twelfths to get $\frac{2}{12}$.

5. $\frac{1}{8}$. If Sandy and her friends ate 7 of the 8 pieces, that means there is 1 piece remaining. Therefore, $\frac{1}{8}$ of the quesadilla still remains.

Measurement and Data

CHAPTER 6

Measurement and Length

We frequently measure items in our lives. You could not buy a shirt, shoes, or a jacket without knowing your measurements. The temperature outside on any given day is a measurement that is on everyone's mind, to decide whether to wear that jacket or not.

When we measure length, we use two different systems: the metric system and the U.S. standard system.

Customary (U.S. Standard) System

The U.S. Standard System is also called the English system because it originated in the United Kingdom. There are four main units of measure in this system:
1 inch (in.) is this length: ....

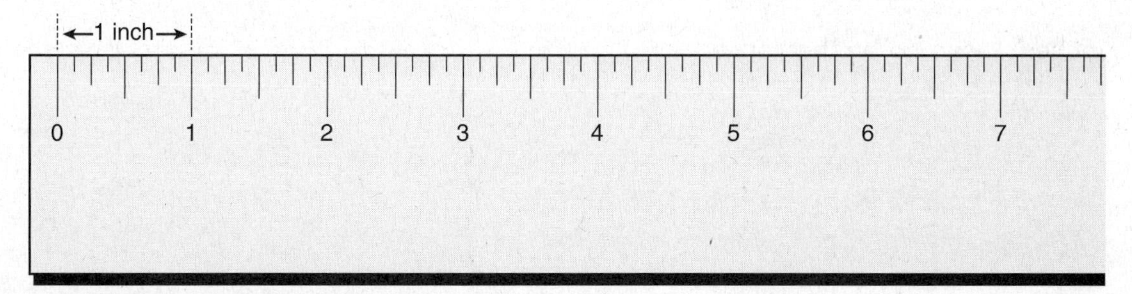

1 foot (ft) = 12 inches
1 yard (yd) = 3 feet or 36 inches
1 mile (mi) = 5,280 feet or 1,760 yards

Use your (customary) ruler to measure these paper clips:

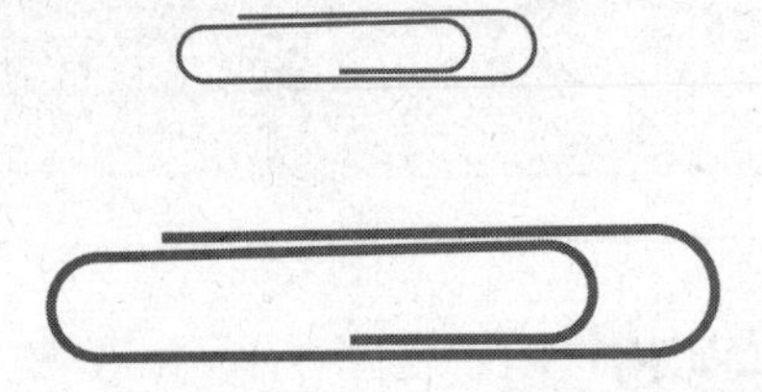

What did you find the length of the small paper clip to be? It should be one inch. Was the larger clip close to 2 inches, but not quite? The second clip should be $1\frac{7}{8}$ inches long.

Try this again, this time on two erasers:

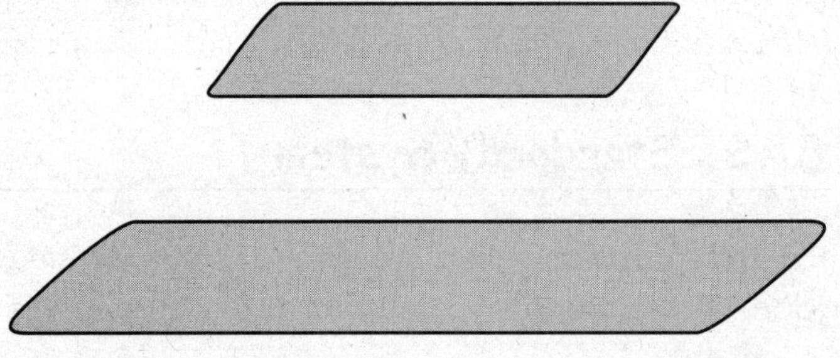

Did you measure the smaller eraser as 2 inches? Was the larger eraser $3\frac{1}{2}$ inches? It should be that length.

Customary (U.S. Standard) System Exercises

1. Measure these stamps with your (customary) ruler. Which one is between $1\frac{1}{2}$ and 2 inches wide?

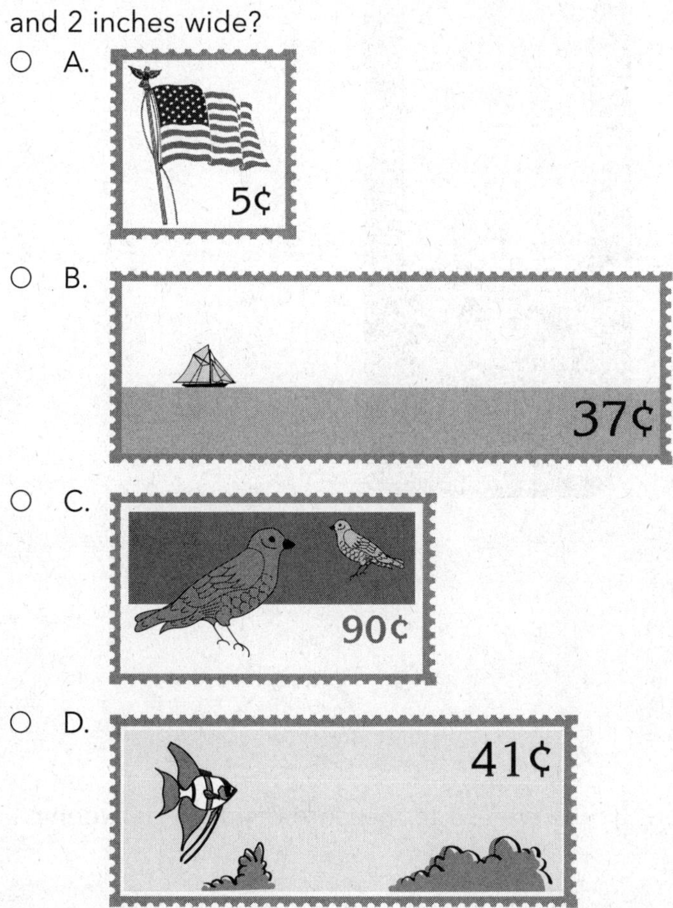

2. Here is an MP3 player. Using your (customary) ruler, tell how long it is.

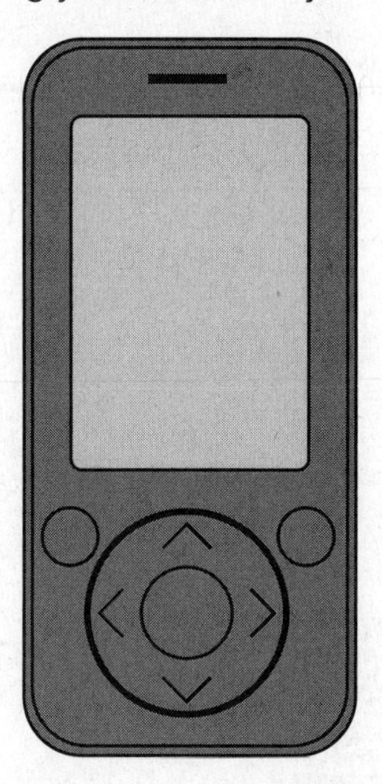

- ○ A. 3 inches
- ○ B. $2\frac{3}{4}$ inches
- ○ C. $3\frac{1}{2}$ inches
- ○ D. 4 inches

3. What unit of measurement would we use to measure distance between cities?

- ○ A. Inches
- ○ B. Feet
- ○ C. Yards
- ○ D. Miles

4. Measure this ship with a (customary) ruler. Pick the pair of numbers that most closely matches the length and height of the ship.

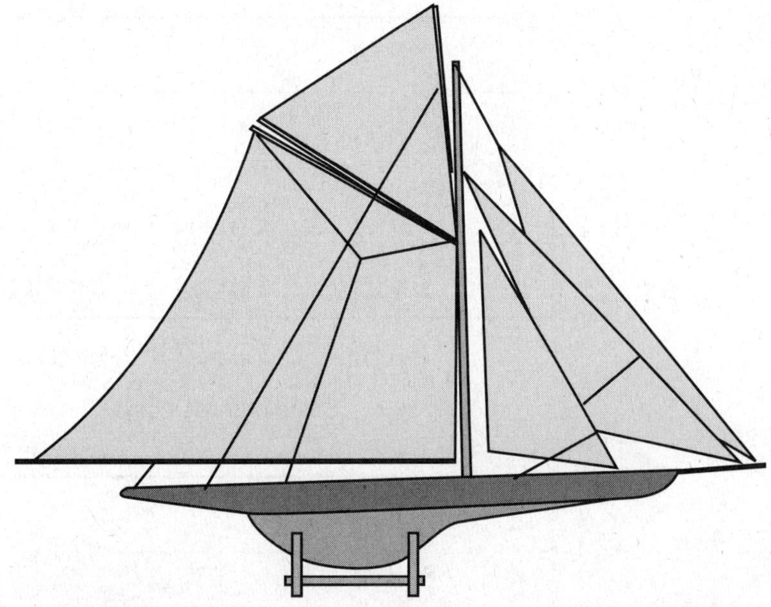

- O A. 2 inches, 4 inches
- O B. 3 inches, 5 inches
- O C. 4 inches, 3 inches
- O D. 5 inches, 2 inches

5. A notebook is one foot high. How many notebooks will make a yard?

- O A. 2
- O B. 3
- O C. 4
- O D. 5

6. Use the measurements below to prepare a line graph showing the sizes of each coin.

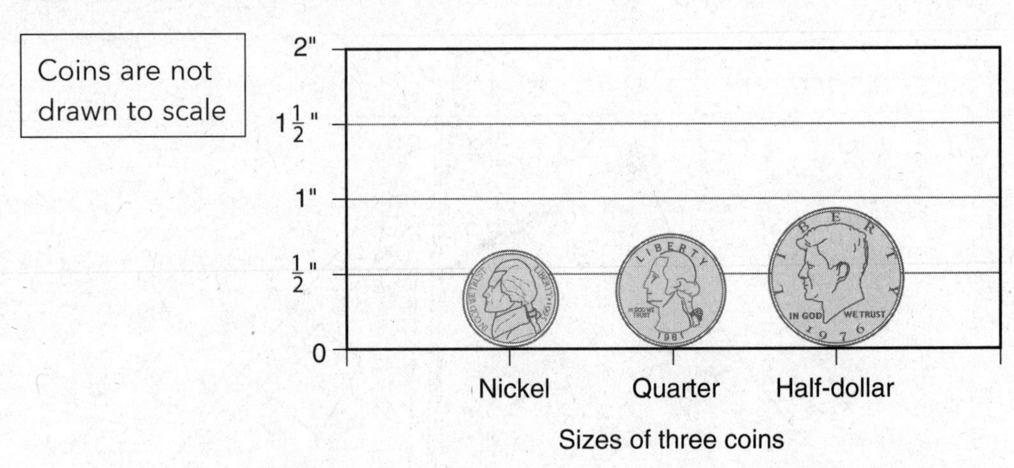

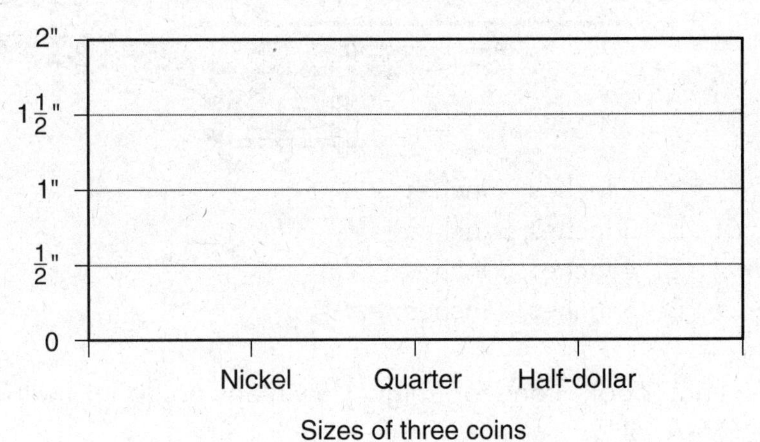

7. Use the measurements below to prepare a line graph showing the sizes of the stamps.

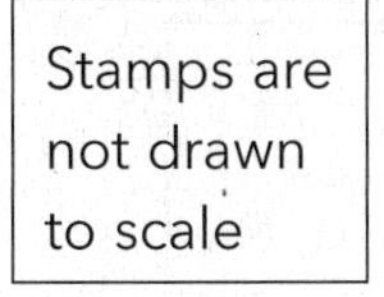

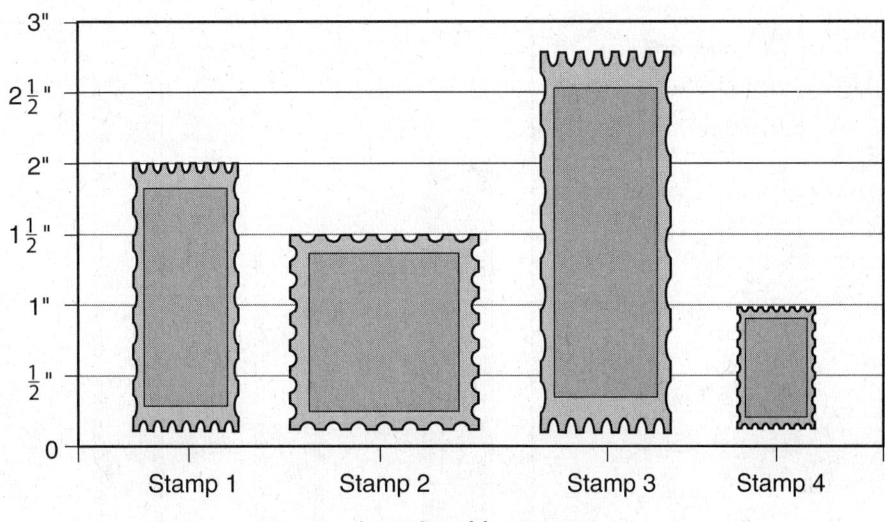

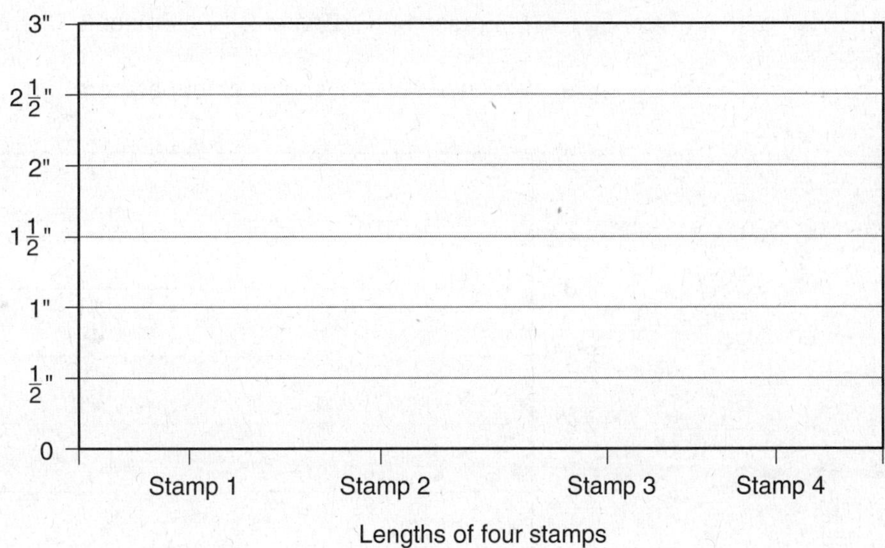

8. Use the measurements below to prepare a line graph showing the sizes of the erasers.

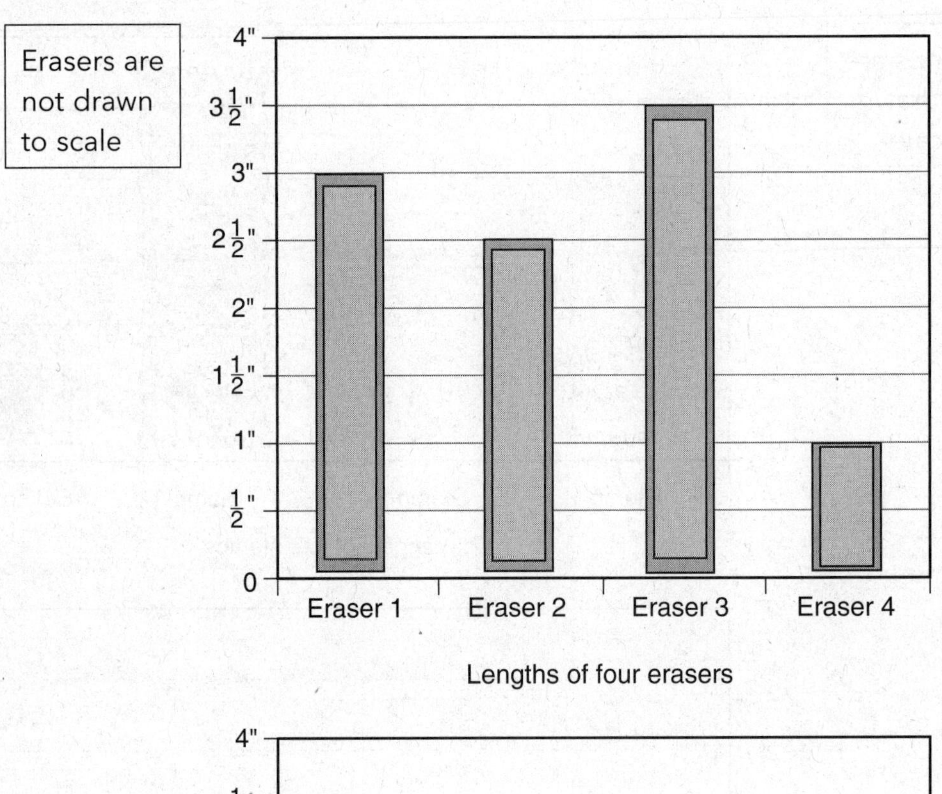

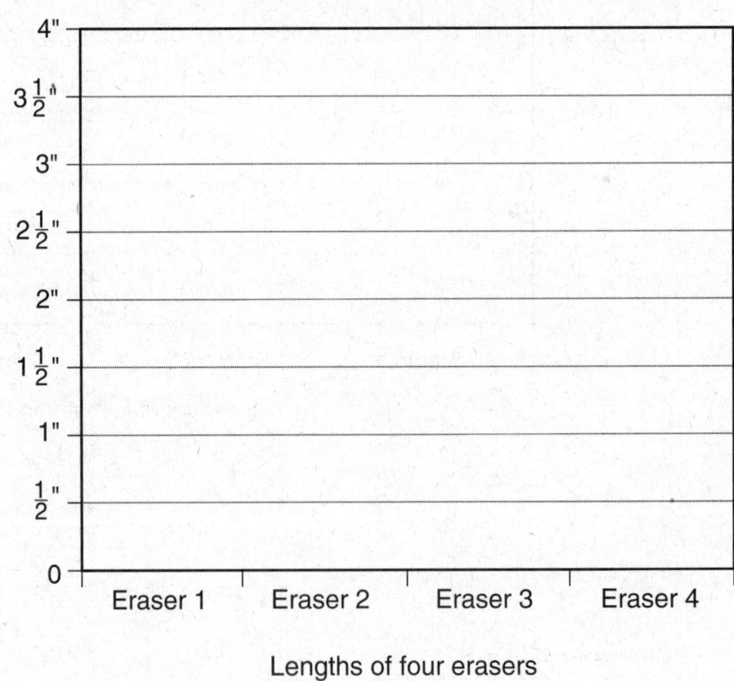

9. Use the measurements below to prepare a line graph showing the sizes of the coins.

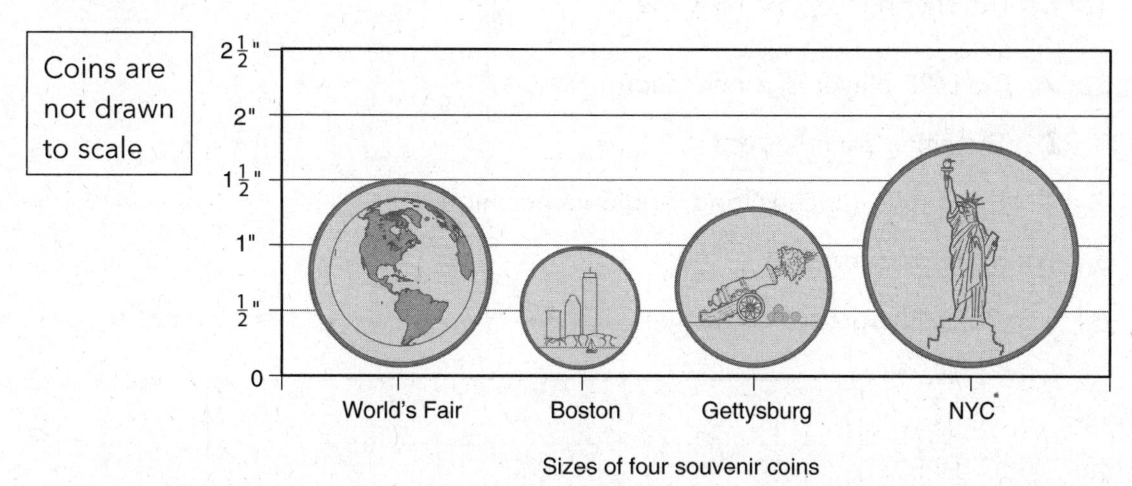

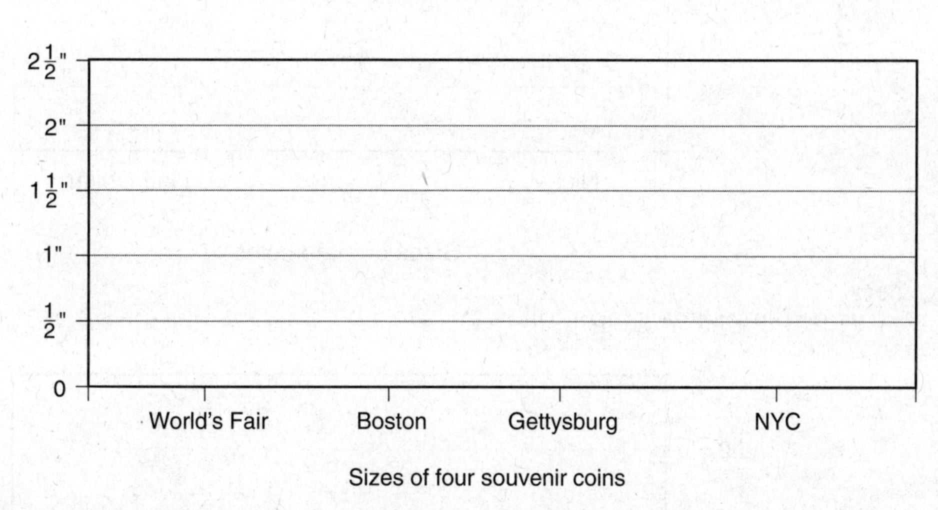

Answers to Customary (U.S. Standard) System Exercises

1. **C.** The stamp is $1\frac{3}{4}$ inches wide.
2. **A.** The MP3 player is 3 inches long.
3. **D.** The cities are miles apart.
4. **C.** The ship is 4 inches long, and 3 inches high.
5. **B.** A yard is 3 feet.
6. The coins line graph follows:

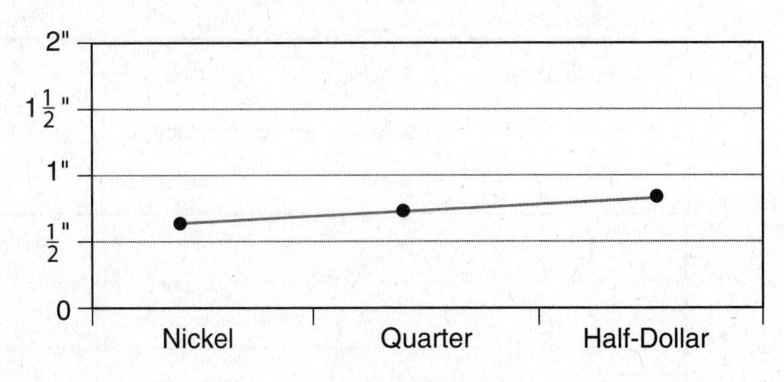

7. The stamps line graph follows:

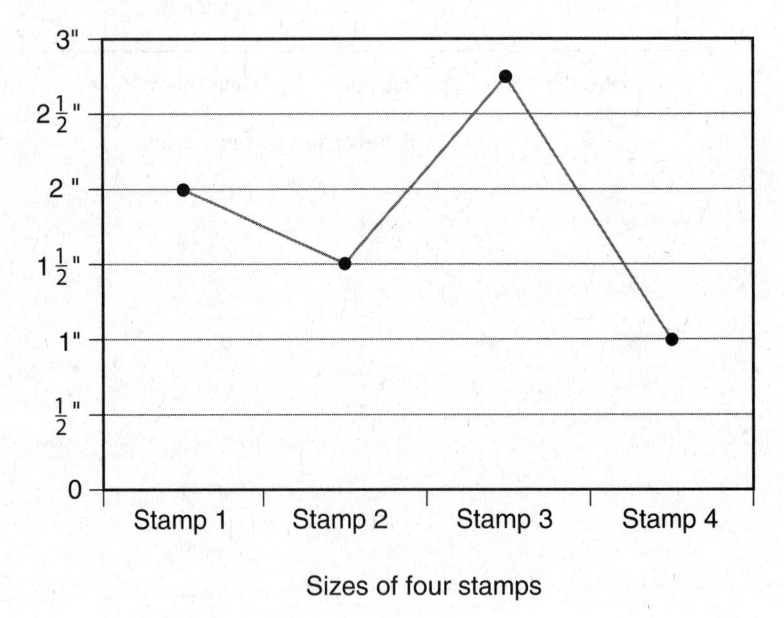

8. The eraser line graph follows:

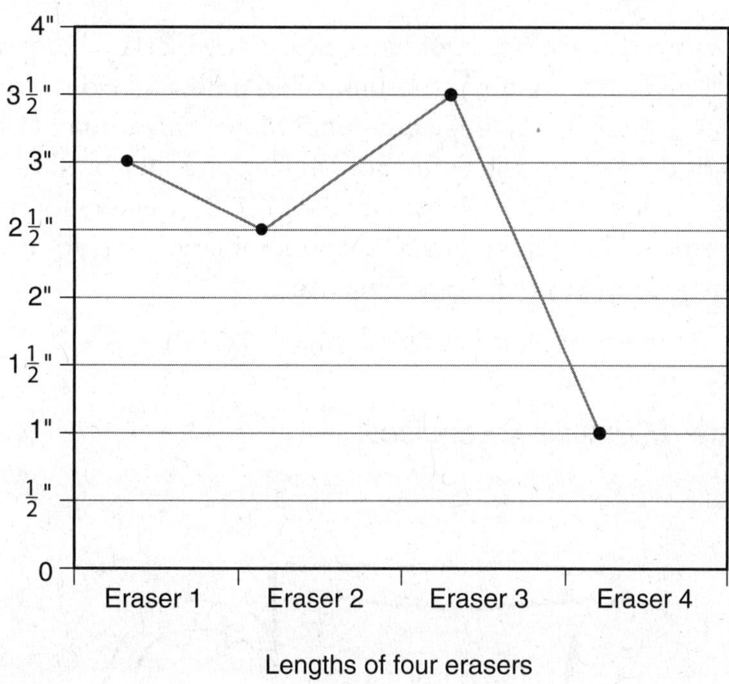

9. The souvenir coins line graph follows:

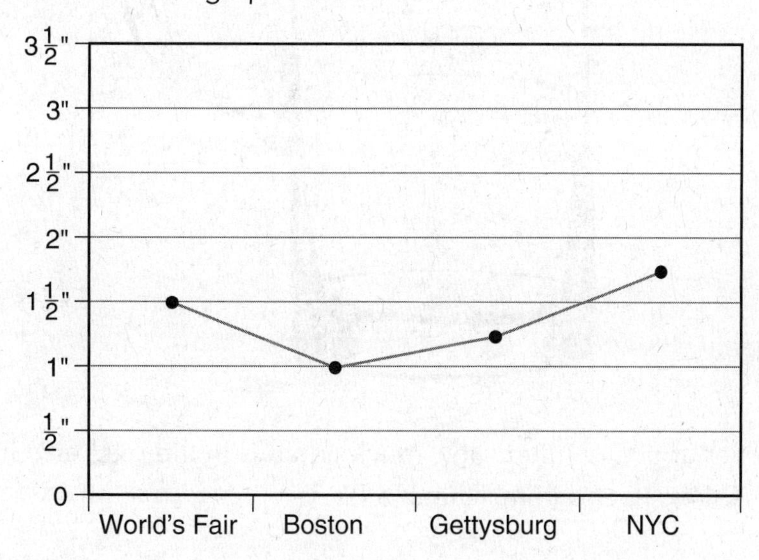

Liquid Volume Measure (Metric)

When we put water or any other liquid in a cup, a bowl, or a pitcher, we measure the amount of liquid with a unit of measure called a **liter** (l). How big is a liter? It's a little more than a quart. It's actually one-tenth more than a quart (1 liter = 1.1 quarts). Most soft drinks and seltzer are sold in liter or 2-liter bottles. There are 1,000 milliliters in a liter. In other words, it takes 1,000 milliliters to make a liter. We measure small amounts of liquid (like a teaspoon of lime juice, for instance) in milliliters. A teaspoon of something is 5 milliliters.

1,000 milliliters (ml) = 1 liter (l)

Liquid Volume Measure Exercises

1.

The pitcher above is 1 liter. How much liquid is in the pitcher? Give your answer in liters (l) and in milliliters (ml)

Answers: .... (l) .... (ml)

2.

The water bottle above is 1 liter. How much liquid is in the water bottle? Give your answer in liters (l) and in milliliters (ml).

Answers: .... (l) .... (ml)

3.

The pitchers above are each 1 liter. How much liquid is in the two pitchers combined? Give your answer in liters (l).

Answer: .... (l)

4.

The three bottles above each are 1 liter. How much liquid is in the three bottles combined? Give your answer in liters (l).

Answer: .... (l)

5.

The three jars above each are 1 liter. How much liquid is in the three jars combined? Give your answer in liters (l).

Answer: .... (l)

6.

The three bottles above each are 1 liter. How much liquid is in the three bottles combined? Give your answer in liters (l).

Answer: .... (l)

7.

The two pitchers above each are 1 liter. How much more liquid is in the first pitcher? Give your answer in liters (l) and in milliliters (ml).

Answers: .... (l) .... (ml)

8.

The three bottles above each are 1 liter. How much more liquid is in the first bottle than in the last two? Give your answer in liters (l) and in milliliters (ml).

Answers: .... (l) .... (ml)

9.

The two jars above each are 1 liter. If we were to multiply by 5, how many liters and milliliters would we have?

Answers: .... (l) .... (ml)

10.

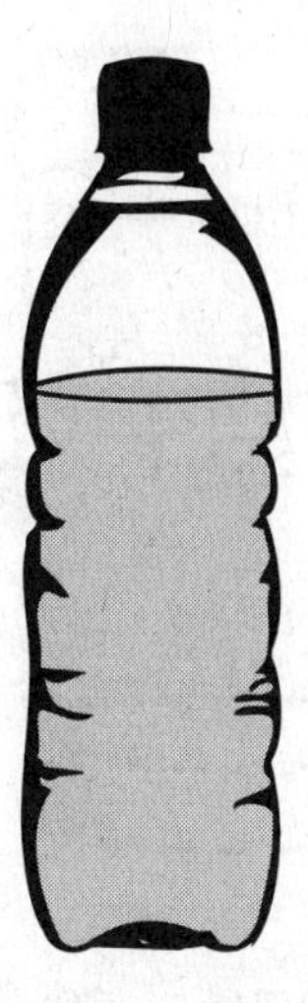

The bottle above is 1 liter. If we were to multiply by 6, how many liters would we have? Give your answer in liters (l).

Answer: .... (l)

11.

The two pitchers above each are 1 liter filled with lemonade. If we are to provide drinks for 8 children, and each child will receive an equal amount, how much lemonade will each child get? Give your answer in liters (l) and in milliliters (ml).

Answers: .... (l) .... (ml)

12.

The three pitchers above each are 1 liter filled with iced tea. If we are to provide drinks for 15 children, and each child will receive an equal amount, how much iced tea will each child get? Give your answer in liters (l) and in milliliters (ml).

Answers: .... (l) .... (ml)

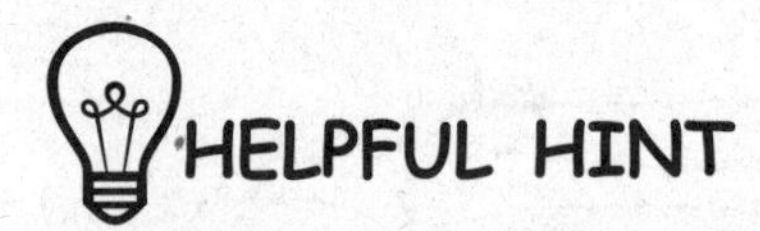

HELPFUL HINT

Estimation of volume capacities—It is useful in various situations to estimate what size bowl or pitcher is needed to hold an amount of liquid.

Answers to Liquid Volume Measure Exercises

1. There is $\frac{1}{2}$ liter, or 500 milliliters in the pitcher.

2. There is $\frac{1}{2}$ liter, or 500 milliliters in the bottle.

3. $\frac{1}{2}+\frac{1}{4}=\frac{3}{4}$. There is $\frac{3}{4}$ liter in the two pitchers combined.

4. $1+1+\frac{1}{4}=2\frac{1}{4}$. There are $2\frac{1}{4}$ liters in the three bottles combined.

5. $\frac{3}{4}+\frac{1}{2}+\frac{1}{4}=1\frac{1}{2}$. There are $1\frac{1}{2}$ liters in the three jars combined.

6. $1+\frac{1}{2}+\frac{3}{4}=2\frac{1}{4}$. There are $2\frac{1}{4}$ liters in the three bottles combined.

7. $1-\frac{1}{4}=\frac{3}{4}$ liter = 750 milliliters. There are 750 milliliters–or $\frac{3}{4}$ liters–more liquid in the first one than in the second.

8. $1-\frac{1}{2}-\frac{1}{4}=\frac{1}{4}$ liter = 250 milliliters. There are 250 milliliters more liquid in the first bottle than in the last two.

9. $1\frac{1}{4}\times 5=6$ liters + 250 milliliters. There would be 6 liters and 250 milliliters in all if you were to multiply the liquid in the two jars by 5.

10. $\frac{3}{4}\times 6=4\frac{1}{2}$. There would be $4\frac{1}{2}$ liters in all if you were to multiply the liquid in the bottle by 6.

11. $2\div 8=\frac{1}{4}$ liter = 250 milliliters. Each child would get $\frac{1}{4}$ liter or 250 milliliters of lemonade. That is about a cup (8 ounces) of lemonade.

12. $3\div 15=\frac{1}{5}$ liter = 200 milliliters. Each child would get $\frac{1}{5}$ liter or 200 milliliters of iced tea. That is about 6 ounces of iced tea.

Volume Estimation Exercises

1. Which of the items above would you use to measure 5 teaspoons of a liquid?

   Answer: --

2. Which of the items above would you use to measure 2 liters of a liquid?

   Answer: --

3. Which of the items above would you use to fill up a bathtub?

Answer: ______________________________

4. Which of the items above would you use on a hike to carry water to drink on the trail?

Answer: ______________________________

Choose the correct liquid measurement for each of the following containers:

5. 3 l or 15 ml?

Answer: ____

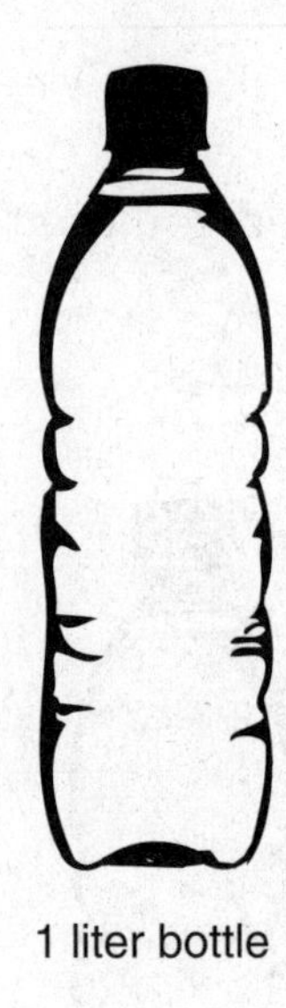

1 liter bottle

6. 500 ml or 7 l?

Answer: ____

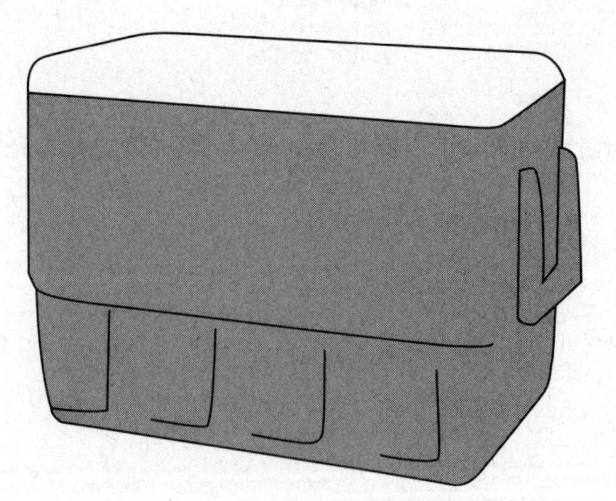

7. 6 l or 50 ml?

Answer: ____

8. 3 l or 60 ml?

Answer: ____________________

Reasoning Exercises

9. Which unit would you use if you were filling up a bucket: milliliters or liters?

Explain your answer: ____________________

____

____

10. Which unit would you use if you were filling up a small cup: milliliters or liters?

Explain your answer: ____________________

____

____

Answers to Volume Estimation Exercises

1. You would use a teaspoon to measure 5 teaspoons of liquid.
2. You would use a water bottle to measure 2 liters of a liquid.
3. You would use a bucket to fill a bathtub.
4. You would use a water bottle on a hike to drink water on the trail.
5. You would use a medicine dose cup to measure out 15 milliliters.
6. You would use a liter bottle to measure out 7 liters.
7. You would use a picnic cooler to measure 6 liters.
8. You would use a child's cup to measure 60 milliliters.

Answers to Reasoning Exercises

9. You would use liters to fill up a bucket because the milliliter is much too small a unit of measure to use when filling up a bucket.

10. You would use milliliters to fill up a small cup because a liter is much too large a unit to fill up a small cup.

Mass Measure (Metric)

We measure mass, the amount of matter in an object, using a unit called the **gram** (g). The gram is quite small. It takes 28 grams to equal one ounce (in the customary system). A **kilogram** (kg) is 1,000 grams, and it is used in many instances for mass. A kilogram is about 2.2 pounds (in the customary system).

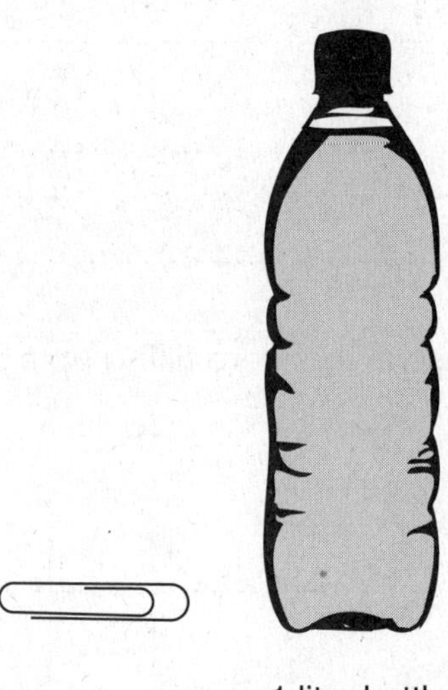

A large paper clip has a mass of about a gram (g).

A liter of water has a mass of a kilogram (kg).

Mass Estimation Exercises

Give the best weight estimate for each of the following:

1.

5 g or 550 g? Answer: ....

2.

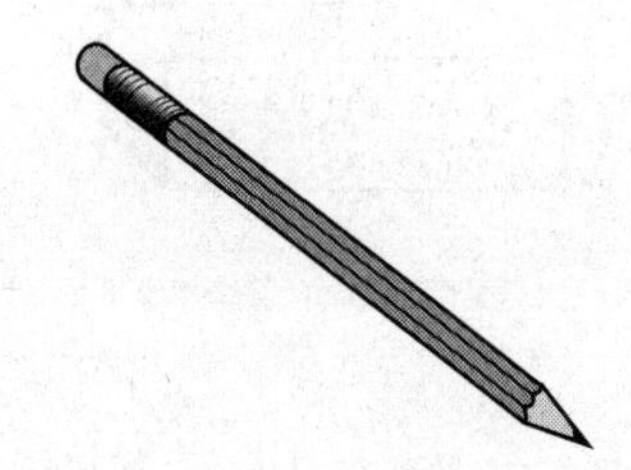

20 g or 2 kg? Answer: ....

3.

4 kg or 150 g? Answer: ....

4.

80 g or 500 g? Answer:

5.

140 g or 2 kg? Answer:

6.

1,000 g or 200 kg? Answer:

7.

140 g or 50 kg? Answer:

8.

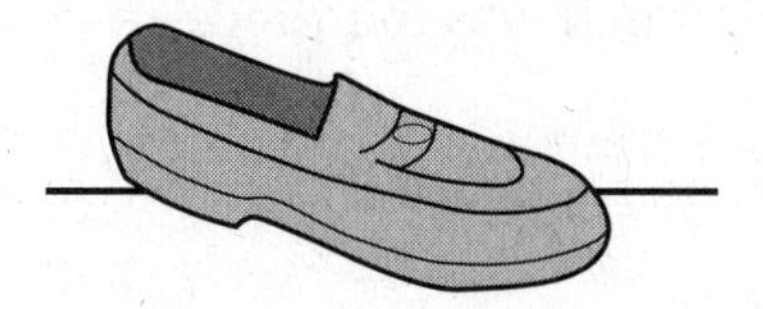

450 g or 3 kg? Answer: ....

Answers to Mass Estimation Exercises

1. Some laptop computers are approximately 550 grams. No laptop is close to 5 grams.
2. A pencil is approximately 20 grams.
3. A cell phone is approximately 150 grams.
4. A squirrel is approximately 500 grams.
5. A brick is approximately 2 kilograms.
6. A motorcycle is approximately 200 kilograms.
7. A coffee cup is approximately 140 grams.
8. A shoe is approximately 450 grams.

Mass Calculation Exercises

1. Filomina had 300 g of sugar in her sugar bowl. She bought 450 g of sugar. How much sugar did she have after she combined the two amounts?

Answer: ....

2. Tenille had 20 kg of potatoes for a picnic she was planning, but she realized she needed more. She went out and bought another 12 kg. How many kg of potatoes did she have then?

Answer: ....

3. Jorge used up 120 g of coffee for a small get-together with his friends. If he originally had 450 g of coffee in his coffee tin, how much does he have now?

Answer: ....

4. A board Edwina was cutting was 3 kg. She cut some of it off, and then the board was 1.8 kg. How much was the piece she cut off?

Answer: ____________________

5. Jessica is a plumber. She needs to cut a piece of pipe to fit into a bathroom fixture. The piece of pipe she needs to cut measures 18 kg. She decides to cut off 9 kg. What is the mass of the pipe now?

Answer: ____________________

6. Danielle is having a party with her friends. She has measured out rice for all the guests. She wants 8 g of rice for each friend, and she has six friends coming. If she makes the same portion for herself, how much rice will she measure out?

Answer: ____________________

7. Coming in from a cold afternoon of sledding, Walter measured out 9 g of hot cocoa mix for each of his eight friends and himself. How many grams of hot cocoa did he measure out?

Answer: ____________________

8. Tim has 30 g of tea in his tea caddy. He wants to make tea for himself and his seven friends. Each cup requires 1.5 g of tea. How much tea will he use? Will there be any tea left in the caddy? How many g will be left?

Answer: ____________________

9. Cosimo went to the store to get some salad for a get-together with his family. He got 78 g of salad. Each member of his family had an equal share of the salad. If the number in his family equals six (including Cosimo), how many g of salad did each member get?

Answer: ____________________

10. The bricks that Harry was working with were just a little too big for the space he needed to place them into. Each brick was 8″ long, but he was able to cut a small slice so it fit into the space. The bricks were 2.5 kg each, but when he cut them to the right size they were 1.9 kg. How much mass did he cut off?

Answer: ____

Answers to Mass Calculation Exercises

1. 300 + 450 = 750. Filomina had 750 grams of sugar when she combined the two amounts.
2. 20 + 12 = 32. Tenille had 32 kilograms of potatoes when she combined the two amounts.
3. 450 − 120 = 330. Jorge had 330 grams of coffee left after he used 120 grams.
4. 3 − 1.8 = 1.2. Edwina cut 1.2 kilograms off the board.
5. 18 − 9 = 9. The pipe was 9 kilograms after Jessica cut 9 kilograms off the pipe.
6. Danielle and her six friends make seven people altogether. 8 × 7 = 56. Danielle will measure out 56 grams of rice for her and her friends.
7. Walter and his eight friends make nine people altogether. 9 × 9 = 81. Walter measured out 81 grams of cocoa for himself and his friends.
8. Tim and his seven friends make eight people altogether. 8 × 1.5 = 12. Tim used 12 grams of tea for himself and his seven friends. That leaves 30 − 12 = 18. 18 grams of tea will be left in the caddy after Tim has made tea for his friends.
9. 78 ÷ 6 = 13. Each member of Cosimo's family had 13 grams of salad.
10. 2.5 − 1.9 = 0.6. Harry cut 0.6 kilograms off each brick to fit them into the space.

Time (Digital Clock)

You use time every day. You get up at a certain time to go to school for a certain number of hours, and usually eat at certain times. **Time** is measured in hours, minutes, and seconds, but we won't be looking at seconds. It's also measured in days, weeks, months, and years, but we will not be talking about those larger units in this section. Hours are measured from midnight with the designation A.M.; hours from noon have the designation P.M. Hours are counted 12 to 1, 2, and so forth up to 11. Minutes are designated 0 to 60.

As an example of time measurement, let's look at a digital clock.

In this clock, the first number (7) gives the hour. The next two numbers (43) give the minutes. The designation P.M. means that it is afternoon or evening. So, it is 7 hours and 43 minutes after noon.

Another example looks at a time before noon.

In this clock, the first number (8) gives the hour, and the next two numbers (23) give the minutes. A.M. means it is before noon, but it really tells us it is 8 hours and 23 minutes *after midnight*. To find out how far it is before noon, you would need to subtract it from 12:00. Remember, though, that hours go only up to 12, and minutes and seconds go only up to 60. So, 11 – 8 = 3, and 60 – 23 = 37. This clock tells us that we are 3 hours and 37 minutes before noon.

Time (Digital Clock) Exercises

1. Matthew woke up and looked at his clock. It read 7:34 a.m. He took a shower, dressed, and went down to make breakfast. When he looked at the clock again, it read 8:46 a.m. How much time had gone by since Matthew had gotten up?

- ○ A. 1 hour, 15 minutes
- ○ B. 2 hours, 20 minutes
- ○ C. 1 hour, 12 minutes
- ○ D. 1 hour, 10 minutes

2. Darlene was working on her homework. She looked at the clock, and it read 8:27 p.m. After studying and writing out some homework sheets, she looked at the clock again, and it read 9:38 p.m. How much time had Darlene worked on her homework?

- O A. 1 hour, 35 minutes
- O B. 1 hour, 11 minutes
- O C. 1 hour, 27 minutes
- O D. 1 hour, 12 minutes

3. Ted coaches wrestling and was giving his wrestlers some drills. He started them at 3:15 p.m. and drilled them for 23 minutes. What time was it then?

- O A. 3:23 p.m.
- O B. 3:33 p.m.
- O C. 3:48 p.m.
- O D. 3:38 p.m.

4. Dan was building a bridge along with seven other boys. They worked together $3\frac{1}{2}$ hours, and they finished the bridge. If you were to spread the hours out, how many hours would that be?

- O A. 28 hours
- O B. $21\frac{1}{2}$ hours
- O C. 23 hours
- O D. 20 hours

5. Joe was studying for his math exam. He started at 7:15 p.m., and studied 2 hours and 35 minutes. What time was it then?

- O A. 9:45 p.m.
- O B. 9:40 p.m.
- O C. 9:50 p.m.
- O D. 9:55 p.m.

Answers to Time (Digital Clock) Exercises

1. **C.** 1 hour, 12 minutes
2. **B.** 1 hour, 11 minutes
3. **D.** 3:38 P.M.
4. **A.** $3\frac{1}{2} \times 8 = 28$
5. **C.** 9:50 P.M.

Time (Analog Clock)

Time is also measured on a clock (sometimes called an analog clock).

In all clocks of this kind, the small hand indicates the hour, and the large hand indicates the minutes. In the clock above, the small hand points to the 3, and the large hand points to the 12, meaning that it is 3 o'clock exactly (no minutes past 3).

In the above clock, the small hand points a little past 3, and the large hand points to 1. All the numbers 1 through 12 indicate 5 minutes, so this clock indicates 5 minutes past 3.

In this clock, the small hand is about a quarter past the 5, and the large hand is pointing to the 3. The large hand indicates 15 minutes past the hour of 5. So this clock indicates a time of 5:15.

In the above clock, the small hand is more than half past the 4, and the large hand is pointing to the 7. The large hand indicates 36 minutes past the hour of 4. So this clock indicates a time of 4:36.

In this clock, the small hand is long past the 6, almost pointing to 7. This means that it is still on the hour of 6. The large hand is in between the 10 and the 11. The large hand indicates 53 minutes past the hour of 6. So this clock indicates a time of 6:53.

Many real-world time problems find the amount of time a person has been on a job. The task is how to find the time that has past from the start to the end of the job.

Example: Jacci bakes cupcakes for a business. She works for 1 hour and 28 minutes baking cupcakes, takes a break, and then bakes for 1 hour and 47 minutes. How long did Jacci bake in all?

To solve this, we add:

$$\begin{array}{rr} 1\text{ hr} & 28\text{ min} \\ +\ 1\text{ hr} & 47\text{ min} \\ \hline 2\text{ hr} & 75\text{ min} \end{array}$$

But there are 60 minutes in an hour, so add 1 hour to the total number of hours and subtract the 60 minutes from the total number of minutes.

Jacci baked 3 hours and 15 minutes.

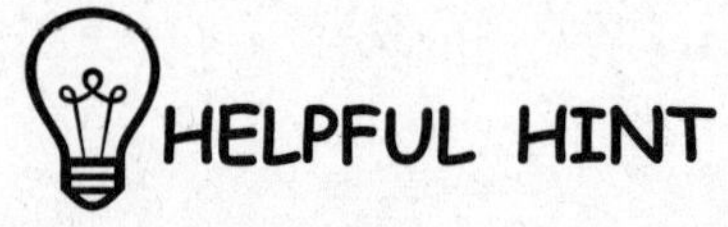

There are 60 minutes in 1 hour, so we carry 1 hour once we get a sum of 60 minutes.

Another example: Allison and Dana own a restaurant. They started cooking pancakes and other delicacies for breakfast at 8: 13 and cooked for 2 hours and 33 minutes. They took a 10 minute break, and then went back and cooked for another 1 hour, 21 minutes. What time was it then?

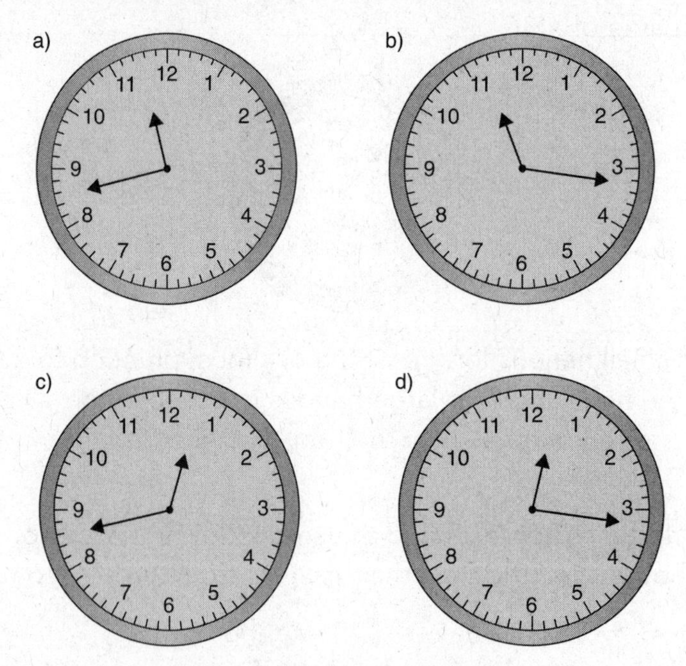

D. Allison and Dana will notice that the time is 12:17.

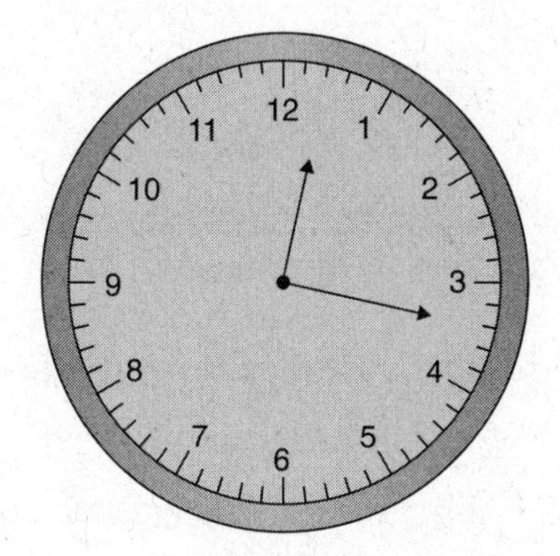

Allison and Dana started at 8:13, worked for 2 hours and 33 minutes:

$$\begin{array}{r} 8{:}13 \\ +\ 2{:}33 \\ \hline 10{:}46 \end{array}$$

Then, they took a 10 minute break:

$$\begin{array}{r} 10{:}46 \\ +\ \ {:}10 \\ \hline 10{:}56 \end{array}$$

Finally, they worked 1 hour and 21 minutes:

$$\begin{array}{r} 10{:}56 \\ +\ 1{:}21 \\ \hline 12{:}17 \end{array}$$

Remember that we carry 60 minutes, not 100 minutes.

At other times, we need to find how much time has elapsed from reading a watch.

Example: Yuan hand delivers important letters by bicycle around Gotham City. He looked at his watch as he started the morning's rounds, and saw that it was 7:36 A.M. He traveled around, delivering his letters, and when he stopped for lunch he saw that the time on his watch was 12:51 P.M. How much time had passed while Yuan delivered letters?

To solve this, we need to subtract:

$$\begin{array}{r} 12 \;\; 51 \\ -\; 7 \;\; 36 \\ \hline 5 \;\; 15 \end{array}$$

Note that we subtract the first time reading from the second time reading; while Yuan delivered letters, 5 hours and 15 minutes passed.

In other problems, we may need to add, as well as subtract.

Example: Yousef has a fresh vegetable business, supplying restaurants in the area. He starts early in the morning, at 6:36 A.M. on his watch. He works all morning, and into the afternoon. When he stops for lunch, he notices that it is 1:27 P.M. How long has Yousef worked today?

To solve this problem, we need to subtract:

$$\begin{array}{r} 12 \;\; 00 \\ -\; 6 \;\; 36 \\ \hline 5 \;\; 24 \end{array}$$

Here we subtracted from noon to get the elapsed time from 6:36 in the morning till noon. Next, we need to add on the time from noon till 1:27:

$$\begin{array}{r} 5 \;\; 24 \\ +\; 1 \;\; 27 \\ \hline 6 \;\; 51 \end{array}$$

Yousef spent 6 hours and 51 minutes working at his business.

Time (Analog Clock) Exercises

1.

Tamara finished her shopping and noticed the clock above. What time did she see? ....

2.

From his desk, Mahmoud glanced at the clock above. What time did he read? ....

3.

Jamie finished his breakfast and glanced at the clock above. What time did he observe? ....

4.

Francois finished his studying and looked at the clock above. What time did he read? ....

5.

Helga finished class and noticed the time. What time did she see?

....

6. Gregory went to the gym and worked out for 1 hour and 20 minutes. He took a break and then worked out another 45 minutes. How long did he work out altogether?

   Gregory worked out for .... hours .... minutes.

7. Tammy delivers packages. On her route she delivered packages for 2 hours and 35 minutes. She took a break for coffee and then delivered packages for another 2 hours and 45 minutes. How long did Tammy deliver packages in all?

   Tammy delivered packages for .... hours .... minutes.

8. On Saturday, Rick was working in his garden. He planted flowers for 1 hour and 45 minutes. He went in for a drink and then came back out and planted for 1 hour and 53 minutes. How long did he plant flowers in all?

   Rick planted flowers for .... hours .... minutes.

9. Alexandra went to her local park in May to paint irises. When she looked at her watch before painting, it read 8:43 A.M. She painted 8 irises and again looked at her watch. It read 11:45 A.M. How long was Alexandra painting flowers?

   Alexandra painted flowers for .... hours .... minutes.

10. Don has a lawn mowing business. When he walked out in the morning, he looked at the clock on the wall, and it read 7:17. He cut five lawns. When he came back to the office in the afternoon, he saw that the clock on the wall read 12:28. How long was Don out of the office mowing lawns?

    Don was mowing lawns for .... hours .... minutes.

Answers to Time (Analog Clock) Exercises

1. 3:15
2. 6:12
3. 8:25
4. 5:54

5. 7:31

6. Gregory worked out 2 hours and 5 minutes. To solve this, we add

$$\begin{array}{rr} 1\text{ hr} & 20\text{ min} \\ + & \underline{45\text{ min}} \\ 1\text{ hr} & 65\text{ min} \end{array}$$

   But there are 60 minutes in an hour, so we add 1 hour to the total number of hours and subtract the 60 minutes from the total number of minutes.

   Answer: 2 hr., 5 min.

7. Tammy delivered packages for 5 hours and 20 minutes. To solve this, we add

$$\begin{array}{rr} 2\text{ hr} & 35\text{ min} \\ +\ \underline{2\text{ hr}} & \underline{45\text{ min}} \\ 4\text{ hr} & 80\text{ min} \end{array}$$

   But, there are 60 minutes in an hour, so we add 1 hour to the total number of hours, and subtract 60 minutes from the total number of minutes.

   Answer: 5 hr., 20 min.

8. Rick planted flowers for 3 hours and 38 minutes. To solve this, we add

$$\begin{array}{rr} 1\text{ hr} & 45\text{ min} \\ +\ \underline{1\text{ hr}} & \underline{53\text{ min}} \\ 2\text{ hr} & 98\text{ min} \end{array}$$

   But, there are 60 minutes in an hour, so we add 1 hour to the total number of hours, and subtract 60 minutes from the total number of minutes:

   Answer: 3 hr., 38 min.

9. Alexandra painted flowers for 3 hours and 2 minutes. To solve this, we subtract:

$$\begin{array}{r} 11:45 \\ -\ \underline{8:43} \\ 3:02 \end{array}$$

10. Don was mowing lawns for 5 hours and 11 minutes. To solve this, we subtract

$$\begin{array}{r} 12:28 \\ -\ \underline{7:17} \\ 5:11 \end{array}$$

Extended Response Questions

Extended response (ER) questions require you to give an answer and then to explain how you got the answer. You could explain your answer using a chart or graph, using pictures, or using words. The examiners of the test award points for the extended response questions on a scale of 0 to 3. They award more points for the more complete explanation. They can award only 1 point to a student who gives the correct answer with no explanation.

Sample Extended Response Questions (for Measurement and Data)

1. Harry is painting the floor of a closet in his house. The dimensions of the floor are shown below.

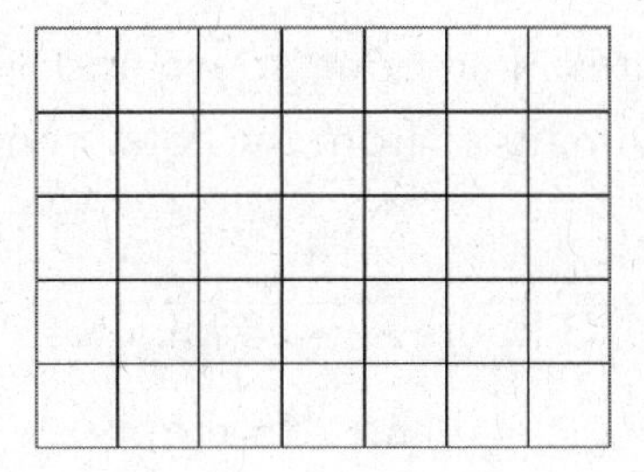

Each square represents a square foot. How would you find out the area of this floor?

Answer to question 1. You could count the number of squares: 35. Another way to solve this would be to multiply the number of feet in the length, which is 7, by the number of feet in the width, which is 5. $7 \times 5 = 35$. A third way might be to add columns of five, seven times, or to add rows of seven, five times. This third way is really multiplication, of course, but it sometimes is used by students who are not completely comfortable multiplying.

2. Effram visited a house of worship in a small village. The building is pictured below.

In this house of worship, identify all the kinds of geometric figures you can.

Tell how many of each kind of figure there is.

Answer to question 2: There are two questions to this problem:

In this picture of a house of worship, there are five kinds of geometric figures: circles, semicircles, triangles, rectangles, and squares

In this picture, there is (are):

- 1 circle (above the main door)
- 3 semicircles (one each above each door)
- 4 triangles (the roof of the house, the roof of the spire, and the two windows on either side of the circle)
- 4 rectangles (the three doors, and the spire)
- 1 square (the house of worship itself)

3. Ted is planning on building a deck in the back of his house. The deck will be 18 meters around the edge, and will be 5 meters long. Tell how wide the deck will be. Ted wants to put little lights every 1 meter around the edge of the deck. How many will he need? How did you get your answer?

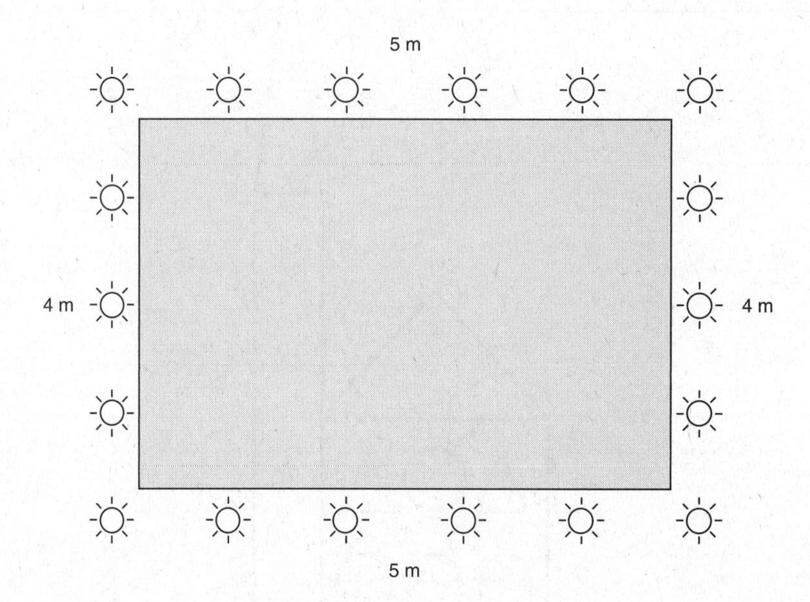

Answer to question 3: There are two questions in this problem:

The deck is rectangular. Since the perimeter is 18 meters around, and the length of the deck is 5 meters, the two lengths together make 10 meters. Then, 18 − 10 = 8. There are 8 meters for the width, but there are two widths, as with the length, so 8 / 2 = 4. The width of the deck is 4 meters. So the dimensions of the deck are: length: 5 meters; width: 4 meters.

We will solve this two ways:

1. To find the number of lights around the edge of the deck, we could start at a corner, go along the length, count one every meter mark, and continue around till we get to the first corner. We can mark those lights at every meter mark: 1, 2, 3, 4, 5, 6, 7, 8, 9, 10, 11, 12, 13, 14, 15, 16, 17, 18. Note that, at light 6, we must then continue down the width of the deck. Similarly, we need to turn the corner at light 10, and at light 15, to get the lights placed around the deck. The total number of lights we need around the deck, then, is 18.

2. Another way to find the number of lights is to note that, along the length, there are six lights, placed at 1 meter intervals. We count them at the meter marks 0, 1, 2, 3, 4, and 5. In a similar fashion, there are six lights along the opposite length. So there are 2 × 6 = 12 lights needed along the top and bottom lengths. Then, along the widths, we only need three lights, because there are lights already at each corner. We need lights at the 3, 6, and 9, and for two widths, that's $2 \times 3 = 6$. So, the total number of lights we need around the deck is $12 + 6 = 18$.

4. How many cubes are needed to make this figure?

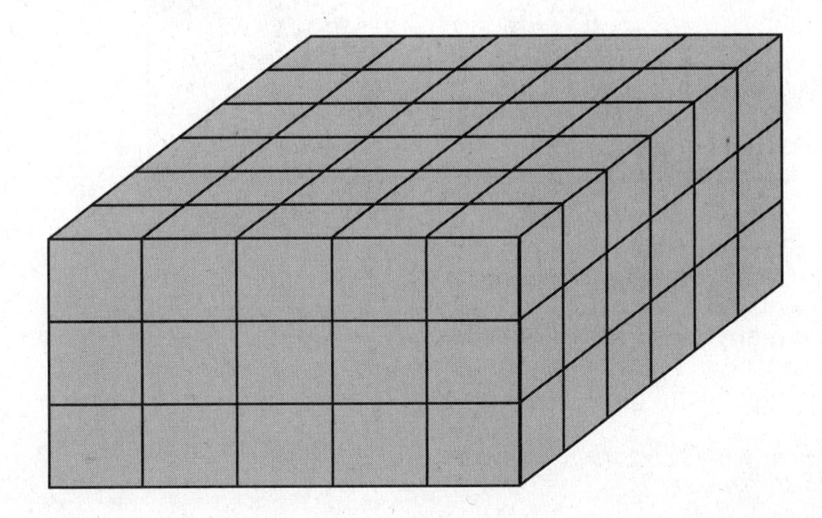

Answer to number 4: This figure is called a rectangular prism. The prism is five cubes long, six cubes wide, and three cubes high. The volume of a rectangular prism like this is length times width times height. So the volume is:

$$V = 5 \times 6 \times 3 = 90$$

Extended Response Question (Measurement and Data) Exercises

1. Look at the ship below.

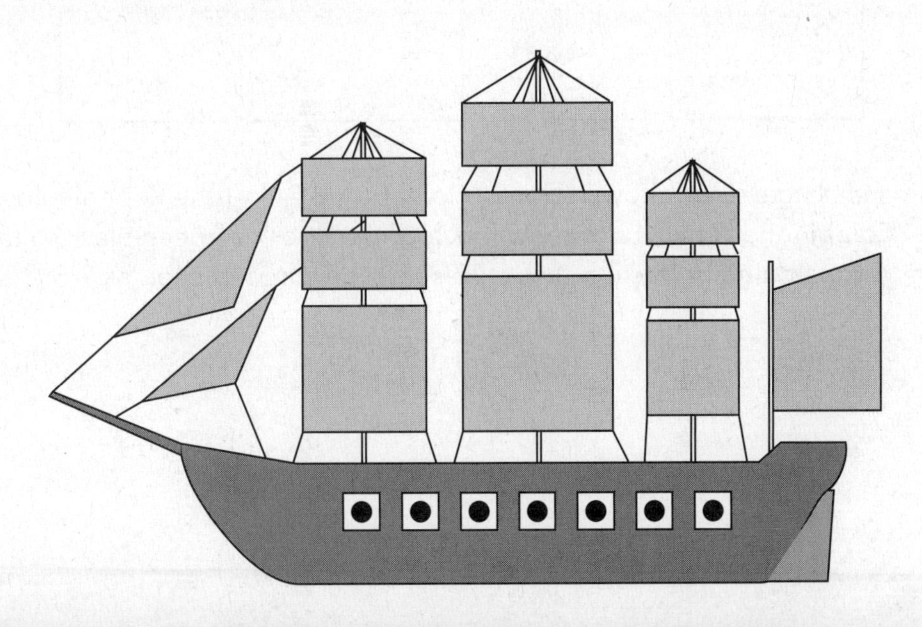

List and name all the geometric figures.

2. Consider the two figures below. (The pyramid below has a square base.)

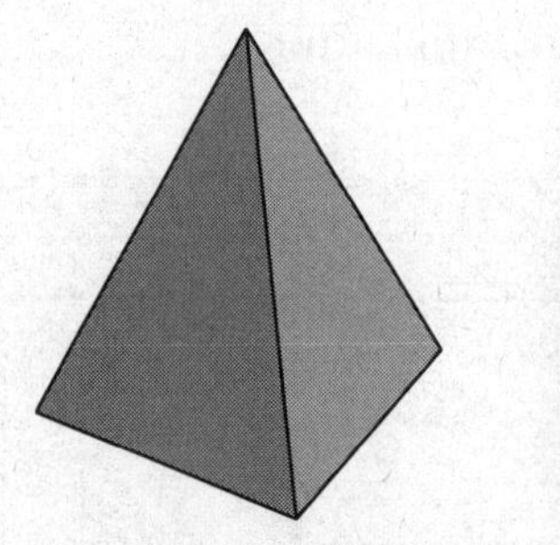

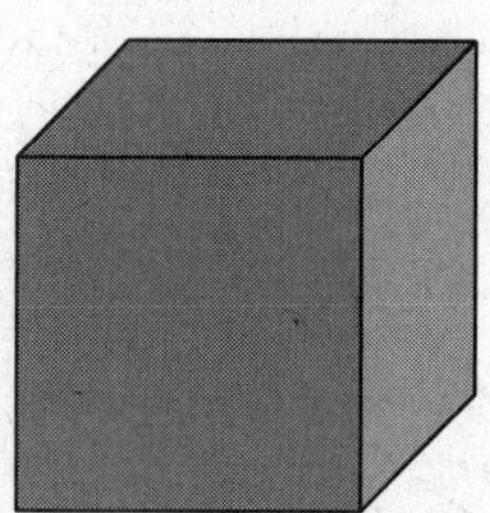

Tell how many edges each figure has.

Tell how many faces each figure has.

Tell one way each figure is the same.

Tell one way each figure is different.

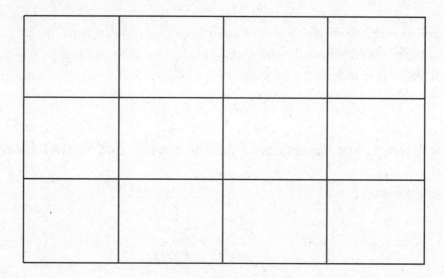

3. Purna is tiling a floor, shown above. She has tiled 2/3 of the floor already.

A. Color in the tiles to show how much of the floor has been tiled, so far.

B. Write a different fraction that represents the same fraction as 2/3:

$$\frac{2}{3} = -$$

C. Explain why the two fractions are equal.

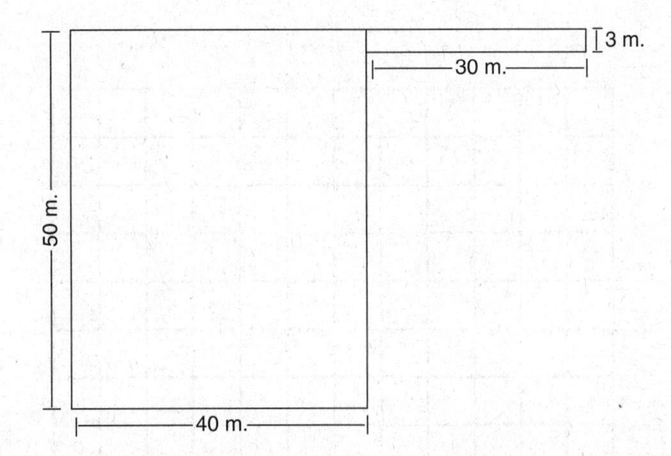

4. Evan works for the South Town Shopping Mall. As shown, it has a central rectangular floor with a hallway leading to an entrance to the parking lot. His job is to carpet the floor and the hallway.

A. Find the area that he is to carpet.

B. He also needs to put a molding at the top of the wall where it meets the ceiling, all around the area of the space and all along the hallway. Find the perimeter of the floor and hallway to find out how much molding he will need.

C. Explain how you calculated the area. Explain how you calculated the perimeter.

5. F. K. Stingray, the Grocery Superstore, wants to fight hunger. So, it sold little square (paper) lunchboxes to customers interested in helping them fight hunger in the town. The paper lunchboxes are one decimeter square (a decimeter is a tenth of a meter).

The drive lasted 5 days. During that time, they sold the following lunchboxes:

Monday	22
Tuesday	18
Wednesday	16
Thursday	21
Friday	13

A. How many paper lunchboxes did F. K. Stingray sell in all?

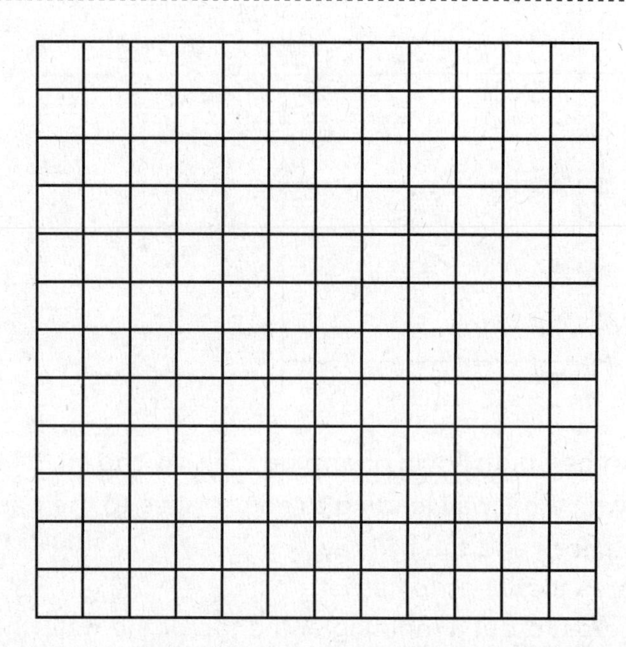

B. Cara works for F. K. Stingray and must decorate one of the store windows. The grid above represents the store window, where each square represents one square decimeter. How could she arrange the lunchboxes in a rectangular array? Fill in boxes to represent how she will decorate the window.

C. Cara made up a rectangular array with 9 lunchboxes in one row. How many rows of lunchboxes will she have? Write a number sentence to show this arrangement.

Answers to Extended Response Question (Measurement and Data) Exercise

1. There are two triangles (the front sails). There are six rectangles (the top two sails on each of the three masts). There are ten squares (the bottom sails on each mast, and the seven cannon ports). There are seven circles (the seven cannons). There is one trapezoid (the sail on the back end, on its own mast).

2. The pyramid has eight edges. The cube has 12 edges. The pyramid has five faces. The cube has six faces.

   a. Similarities: The pyramid and the cube are both approximately the same size. The pyramid and cube are both three-dimensional figures. The pyramid and the cube both have only straight edges. The pyramid and cube both have only flat faces.

b. Differences: The pyramid and cube each have a different number of faces, and a different number of edges.

3. ***A.***

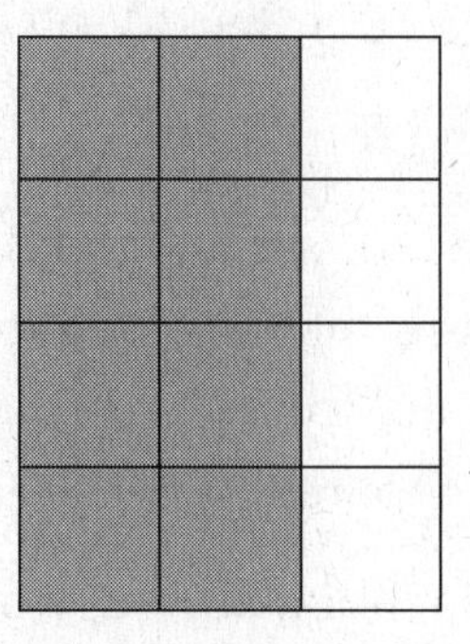

B.

$$\frac{2}{3} = \frac{4}{6} \text{ or } \frac{6}{9}$$

C. To explain part B, note that if we divide the same number from the numerator and the denominator, we'll get the same fraction as the original:

$$\frac{4}{6} = \frac{(2)2}{(2)3} = \frac{2}{3}$$

or

$$\frac{6}{9} = \frac{(3)2}{(3)3} = \frac{2}{3}$$

Note also that there are many other fractions that are equal to the fraction of the floor that Purna has tiled.

4. ***A.*** The space can be divided into two rectangular areas, the main floor and the hallway. The floor is 40 m × 50 m, so the area of the floor is

$$40 \times 50 = 2{,}000 \text{ m}^2$$

Then, the hallway is 30 m × 3 m, so that area is

$$30 \times 3 = 90 \text{ m}^2$$

So, the total area that Evan needs to carpet is: 2,000 + 90 = 2,090 m^2.

B. To find the perimeter, we add all the sides of the space. Going around the floor, we have a 40 m wall, then 50 m and then the floor and the hallway, 40 + 30 = 70 (because it's the sum of the shorter wall and the hallway).

Then, we have 3 m for the end of the hallway, followed by 30 m of the other side of the hallway. The last length is 47 m because it is 50 m minus the 3 m representing the width of the hallway. The perimeter of the whole space is

$$40 + 50 + 70 + 3 + 30 + 47 = 240 \text{ m}$$

Evan will need to get molding for 240 m.

C. To find the area of the space, we divided the space into two separate rectangles. Then we calculated the area of each rectangle, and finally we add the two areas to get the area of the whole space. To find the perimeter, we mark off the length of each separate wall and add them.

5. **A.** To find the number of lunchboxes sold, we add the number sold each of the 5 days:

$$22 + 18 + 16 + 21 + 13 = 90$$

F. K. Stingray sold 90 lunchboxes.

B. We need to model how to arrange 90 square decimeters (dm^2). 90 can be the product of 10×9, or 15×6. Since the window is 12×12 dm^2, we cannot arrange 15×6 in the window. Now, 10×9 could be displayed two ways: either as a 10×9 array, or as a 9×10 array, as shown:

C. If Cara arranged the paper lunchboxes into rows 9 lunchboxes long, she would have 10 rows. The number sentence that would describe this arrangement is 9×10.

Data Analysis

We use data analysis all the time to order our world. Sometimes teachers arrange students in their classes in alphabetical order, and sometimes they order them by height. At the ice cream shop, you can find out how many possibilities you have with the ice cream flavors available, the cone types, and the toppings, using discrete mathematics. In the remainder of this chapter, we'll be looking at these topics.

When you collect information, such as the heights of all your classmates, or how many have red shirts and how many have blue shirts, you are collecting **data**. When you put that data in a chart or a graph of some kind and then make judgments based on the graph, that is **data analysis**. Let us look at a simple example of this.

In a class of 22 students, it was found that:

14 students have brown eyes,
6 students have blue eyes, and
2 students have green eyes.

To make these data easier to understand (or see), it can be put in the form of a pictograph:

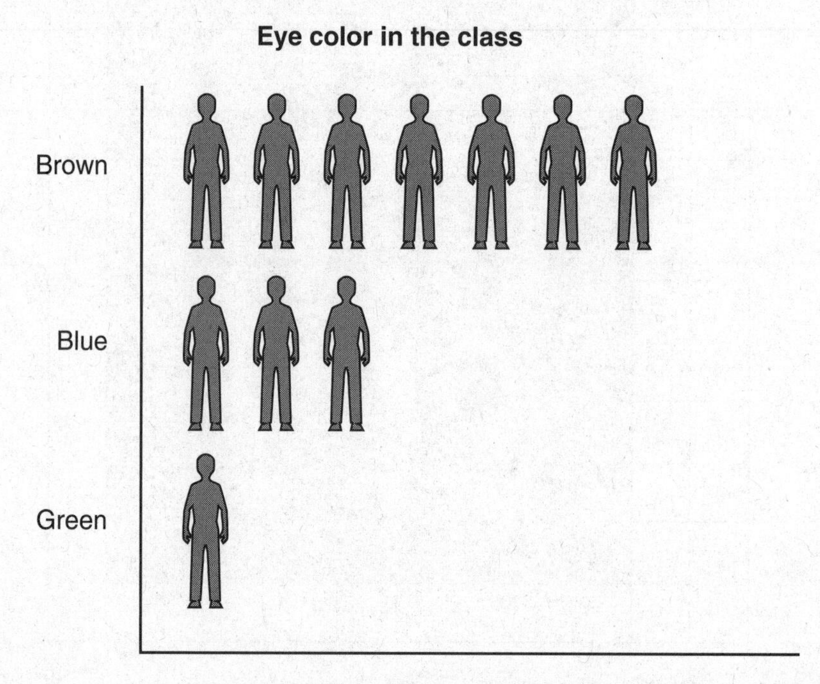

Pictographs often will have pictures stand for the data they represent. The pictures often will be a picture of the thing they represent. So, in this pictograph, there are pictures of students, and each picture of a student stands for two students. In national pictographs, one person might stand for a thousand (or even a million) people.

These data can also be put in the form of a table:

Brown	14
Blue	6
Green	2

The data can also be in the form of a bar graph:

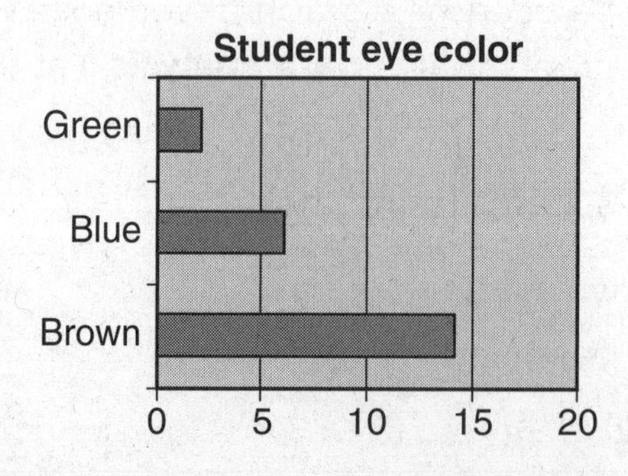

Now we could go the other way. We could look at a graph and make observations based on that graph. Consider this graph.

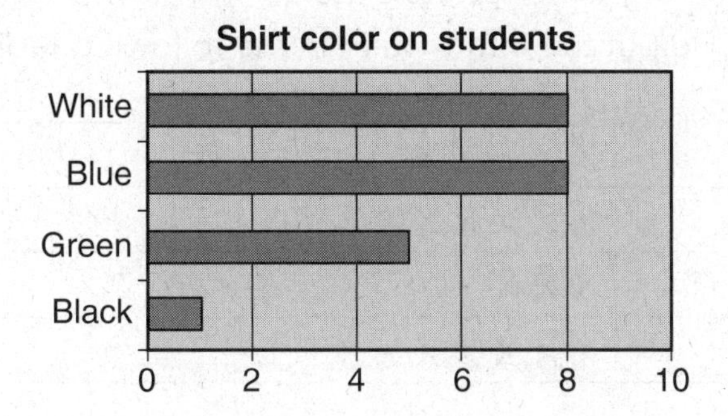

In this example, how many students would you say:

Are wearing black shirts?

Are wearing green shirts?

Are wearing blue shirts?

Are wearing white shirts?

There is one student wearing a black shirt.

There are five students wearing green shirts.

There are eight students wearing blue shirts.

There are eight students wearing white shirts.

We can make certain calculations based on these data as well. For example:

How many more students are wearing blue than black?

Since there are eight students who are wearing blue and one student who is wearing black, there are $8 - 1 = 7$. Seven more students are wearing blue than are wearing black.

How many students are in the class?

Since there are eight students wearing white, eight students wearing blue, five students wearing green, and one student wearing black, then we have

$$8 + 8 + 5 + 1 = 22$$

There are 22 students in the class.

Data Analysis Exercises

1. Emelio was ordering pizza for the class party, and wrote down what all the students wanted. He put the information in the following table:

Plain	X X X X X
Pepperoni	X X X X X X X X X X
Mushrooms	X X X
Sausage	X X X X X X X X

Based on this information, which of the following is true?

- ○ A. There are more students who like sausage than pepperoni.
- ○ B. There are fewer students who like mushrooms than plain.
- ○ C. There are more students who like plain than sausage.
- ○ D. There are more students who like mushroom than sausage.

2. Kara drew up a chart that represents the number of students born in the different months of the year:

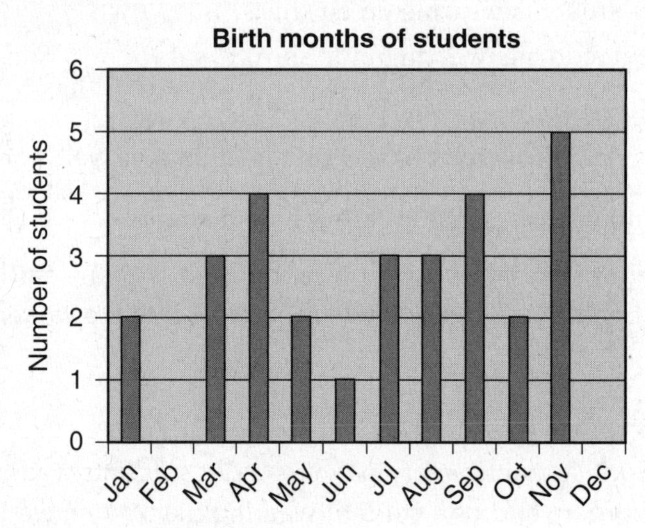

a. How many students were born in the first half of the year (January to June)?

- ○ A. 12
- ○ B. 11
- ○ C. 14
- ○ D. 15

b. How many more students were born in July than in June?

---

3. Warren is a traffic officer and was looking at the cars driving down Gravel Rd. He wrote his findings in a chart:

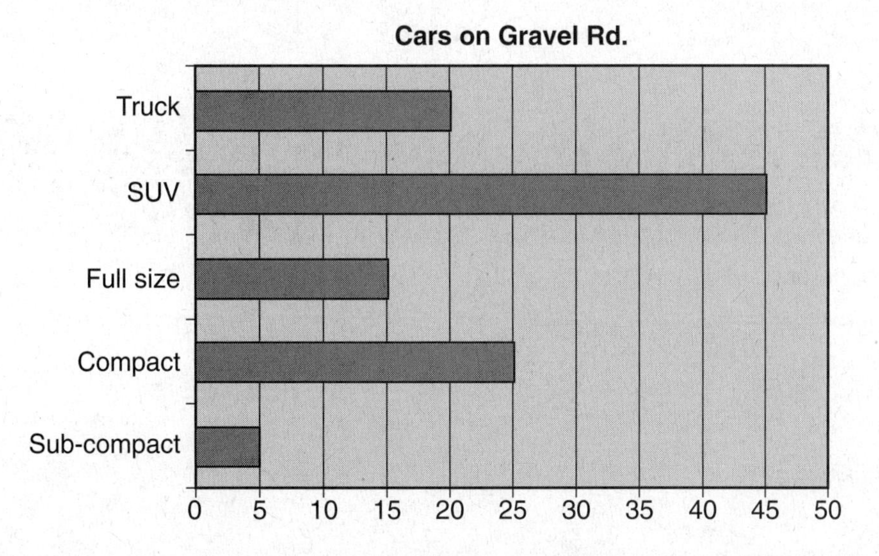

Which statement is true, based on the data?

- ○ A. Trucks are the most popular vehicles on Gravel Rd.
- ○ B. Compacts are the least popular vehicles on Gravel Rd.
- ○ C. Full-size cars are the least popular vehicles on Gravel Rd.
- ○ D. The SUVs are the most popular vehicles on Gravel Rd.

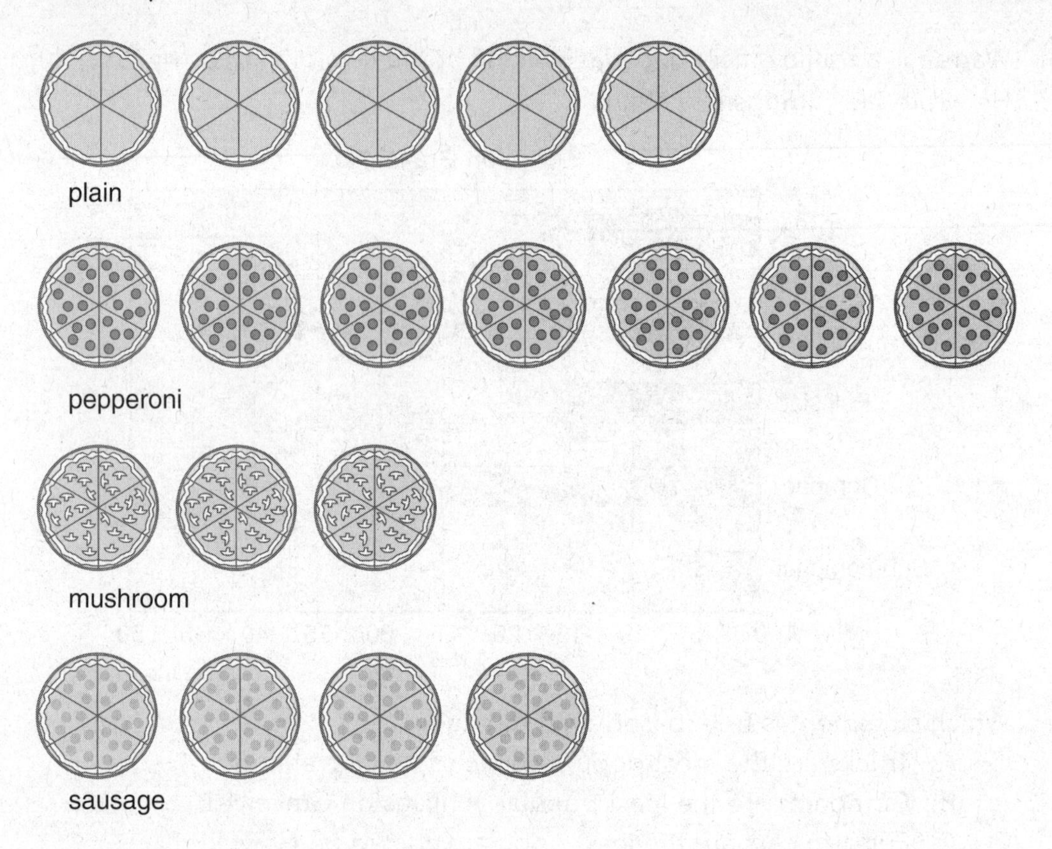

4. The above figure represents the pizzas ordered by the Walker St. School for their Field Day. As the legend says, each picture stands for four pizzas. Using this pictograph, please answer the following questions:

A. How many plain pizzas did the Walker St. School order? ....

B. How many pepperoni pizzas did the Walker St. School order? ....

C. How many more plain pizzas than mushroom pizzas did the Walker St. School order? ....

D. How many more pepperoni pizzas than sausage pizzas did the Walker St. School order? ....

E. How many pizzas altogether did the Walker St. School order? ....

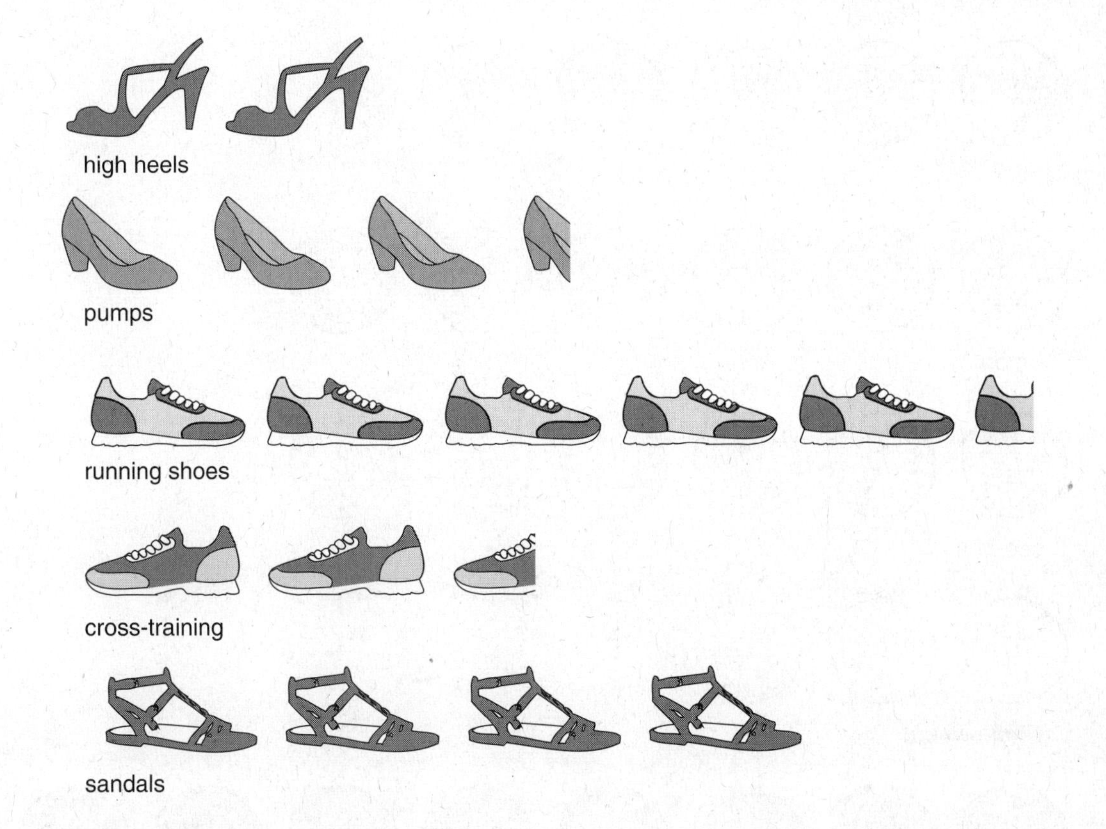

5. The above figure represents the different amounts and types of shoes in the ladies shoe department of Sibley's department store.

A. How many pairs of shoes are in the ladies shoe department? ________

B. How many pairs of running shoes and cross-training shoes combined are there? ________

C. How many more pairs of pumps than high heeled shoes are in the ladies shoe department? ________

D. How many fewer pairs of sandals than running shoes are there? ________

E. If we consider high heeled and pump shoes to be dress shoes, and running shoes, cross-training shoes, and sandals to be casual shoes, how many fewer pairs of dress shoes than casual shoes are in the ladies shoe department? ________

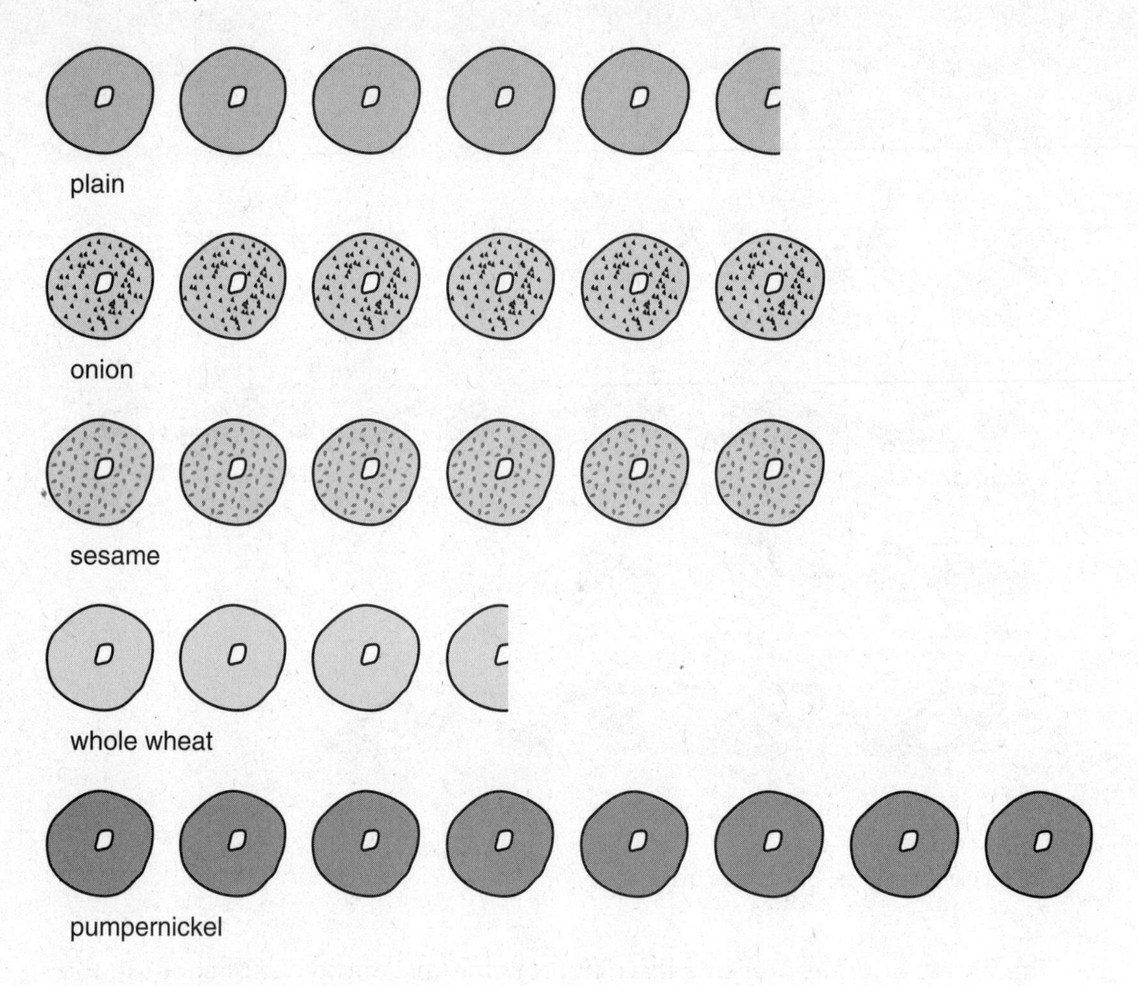

Bengal's Bagels. Each picture represents ten bagels

6. The pictograph above represents the number of bagels sold by Bengal's Bagels on one cold November day.

A. How many onion bagels did Bengal's sell? _______

B. How many more pumpernickel than plain bagels were sold? _______

C. How many more onion than whole wheat bagels did Bengal's sell? _______

D. How many more onion and sesame bagels than plain bagels did Bengal's sell? _______

E. What was the total number of bagels sold by Bengal's that cold November day? _______

7. Giselle's Pants Depot sells just three kinds of pants: sweatpants, jeans, and cargo pants. When she opened up on the first of April, Giselle had 30 pairs of sweatpants, 60 pairs of jeans, and 50 pairs of cargo pants. Draw a bar graph representing these amounts of pants on the graph below:

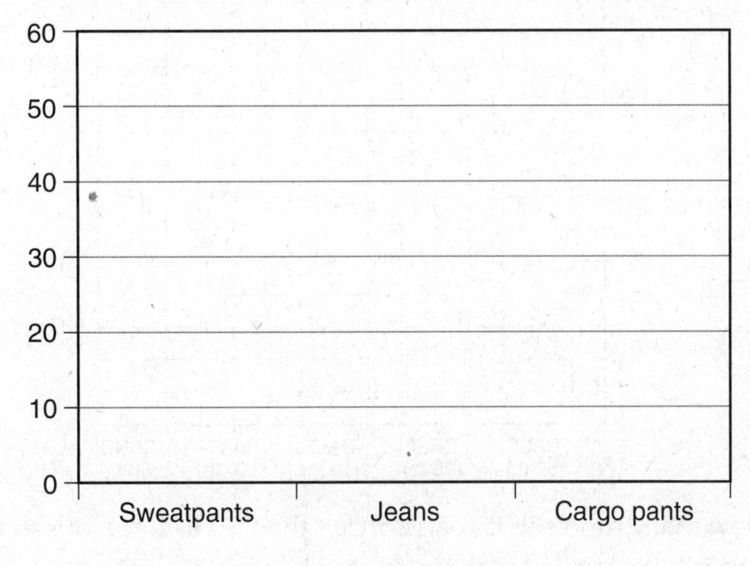

8. Ilya is manager of Handyman Hardware. For the winter season, he ordered salt and sand to sell to homeowners and contractors. He ordered salt and sand in the following amounts:

150 bags of 50 pounds of salt
250 bags of 25 pounds of salt
300 plastic containers of 10 pounds of salt
50 bags of 100 pounds of sand
75 bags of 50 pounds of sand

Complete the following chart showing the number and types of salt and sand ordered by Handyman Hardware.

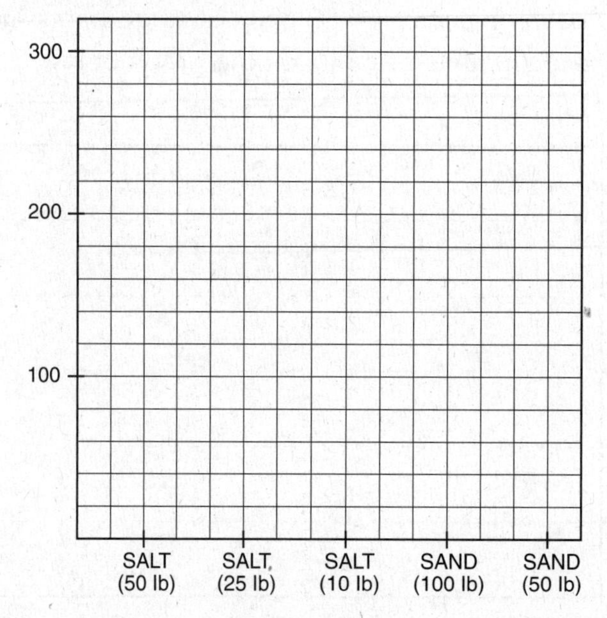

A. How many more 25 lb bags of salt than 50 lb bags of salt did Ilya order?

B. How many fewer 100 lb bags of sand than 50 lb bags of sand did Ilya order?

C. How many bags of sand did Ilya order in all?

D. How many bags and containers of salt did Ilya order in all?

E. How many more bags and containers of salt than bags of sand did Ilya order?

---

9. The Biotech Club put on a fundraiser for a charity they wanted to support. Stephanie is the president, and she tallied up the amount of snacks the Biotech Club sold that day:

35 cupcakes
50 cakepops
40 brownies
25 candy bars
40 bags of chips
40 bottles of drinks

Fill out the following chart showing all the snacks the Biotech Club sold.

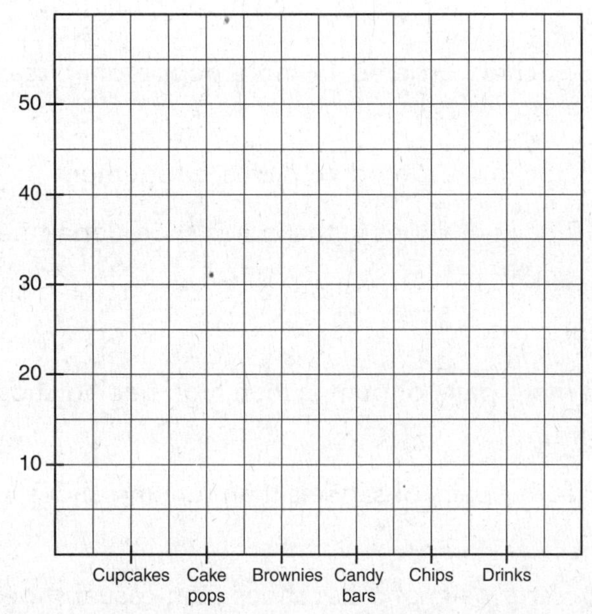

A. How many sweet items (cupcakes, cakepops, brownies, and candy bars) did the Biotech Club sell?

B. How many fewer candy bars than drinks did the Biotech Club sell?

C. How many more chips than candy bars did the Biotech Club sell?

D. How many cupcakes and cakepops did the Biotech Club sell?

E. How many more cakepops than drinks did the Biotech Club sell?

Answers to Data Analysis Exercises

1. **B.** There are fewer students who like mushrooms than plain.
2. **A. a.** 12 students were born in the first half of the year.

   b. There were 2 more students born in July than in June.
3. **D.** The SUVs are the most popular vehicles on Gravel Rd.
4. **A.** The Walker St. School ordered 20 plain pizzas.

   B. The Walker St. School ordered 28 pepperoni pizzas.

   C. The Walker St. School ordered eight more plain pizzas than mushroom pizzas.

   D. The Walker St. School ordered 12 more pepperoni pizzas than sausage pizzas.

   E. The Walker St. School ordered 76 pizzas, altogether.
5. **A.** There are 210 pairs of shoes in the ladies shoe department.

   B. There are 96 pairs of running shoes and cross-training shoes in the ladies shoe department.

   C. There are 18 more pairs of pumps than high heeled shoes in the ladies shoe department.

   D. There are 18 fewer pairs of sandals than running shoes in the ladies shoe department.

   E. There are 78 fewer pairs of dress shoes than casual shoes in the ladies shoe department.
6. **A.** Bengal's Bagels sold 60 onion bagels.

   B. Bengal's Bagels sold 25 more pumpernickel bagels than plain bagels.

   C. Bengal's Bagels sold 25 more onion bagels than whole wheat bagels.

   D. Bengal's Bagels sold 65 more onion and sesame bagels than plain bagels.

   E. Bengal's Bagels sold 290 bagels in all that cold November day.

7.

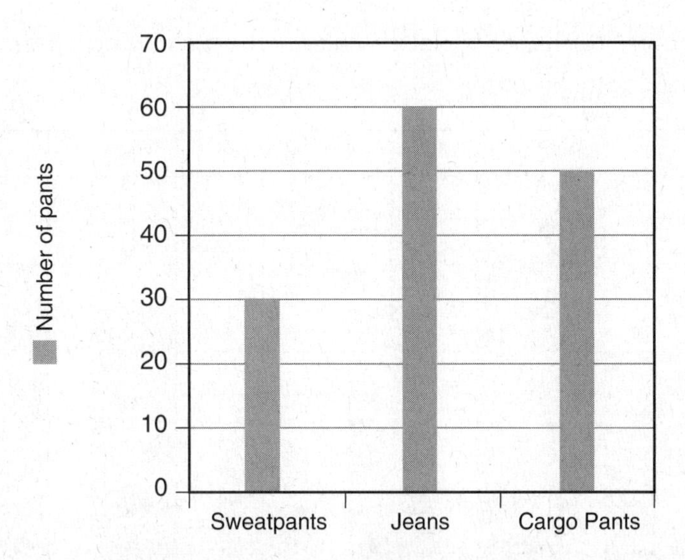

8. If Ilya were to make a chart showing salt and sand, it would look like this:

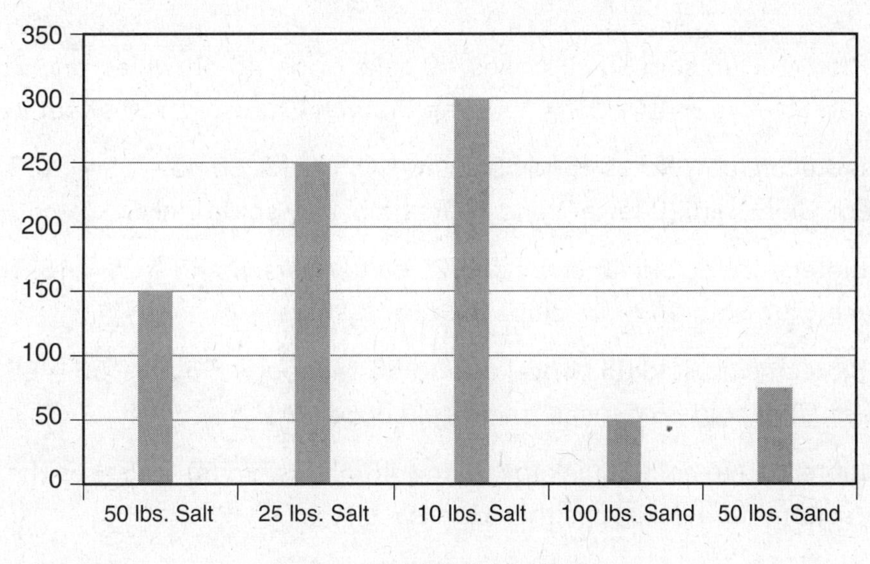

A. Ilya ordered 250 25 lb bags of salt, and 150 50 lb bags of salt. So 250 – 150 = 100. Ilya ordered 100 more bags of 25 lb of salt than the 50 lb bags of salt.

B. Ilya ordered 75 50 lb bags of sand and 50 100 lb bags of sand. So 75 – 50 = 25. Ilya ordered 25 fewer bags of 100 lb of sand than the 50 lb bags of sand.

C. 75 + 50 = 125. Ilya ordered 125 bags of sand in all.

D. 250 + 150 + 300 = 700. Ilya ordered 700 bags and containers of salt in all.

E. 700 – 125 = 575. Ilya ordered 575 more bags and containers of salt than bags of sand.

9. Biotech Club. If Stephanie were to make a chart showing all the snacks she sold, it would look like this:

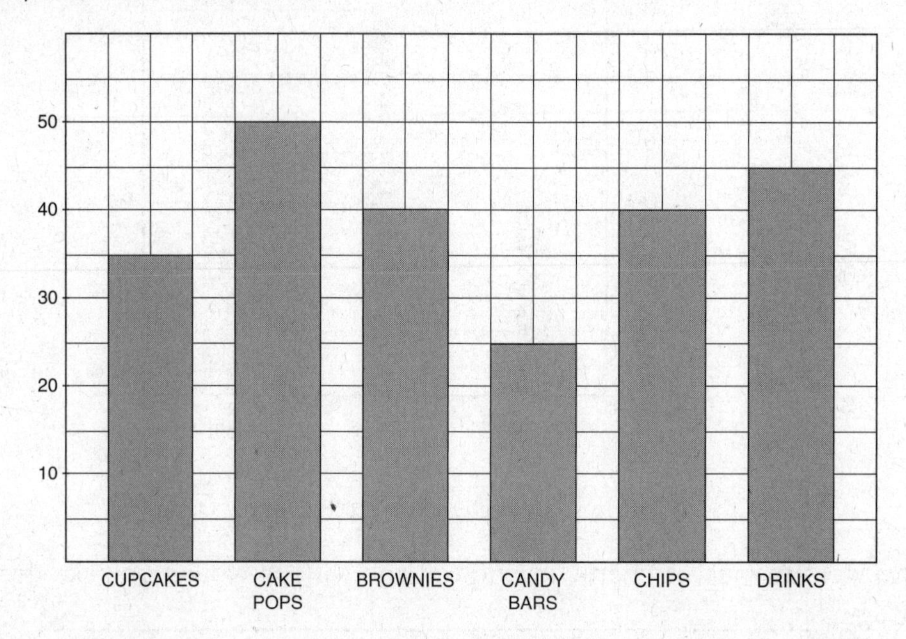

A. The Biotech club sold 35 cupcakes, 50 cake pops, 40 brownies, and 25 candy bars. So 35 + 50 + 40 + 25 = 150. The Biotech Club sold 150 sweet items.

B. The Biotech club sold 25 candy bars and 45 drinks. So 45 – 25 = 20. The Biotech Club sold 20 fewer candy bars than they sold drinks.

C. The Biotech Club sold 40 chips and 25 candy bars. So 40 – 25 = 15. The Biotech Club sold 15 more chips than candy bars.

D. The Biotech club sold 35 cupcakes and 50 cakepops. So 35 + 50 = 85. The Biotech Club sold 85 cupcakes and cakepops.

E. The Biotech club sold 50 cakepops and 45 drinks. So 50 – 45 = 5. The Biotech Club sold 5 more cakepops than drinks.

Geometry

CHAPTER 7

Geometric shapes are everywhere in our world. We play with lots of different sized balls, which are spheres. Sugar cubes are just that: cubes. We write on pieces of paper, which most of the time are shaped like rectangles. We eat pie, which is shaped like a circle. Along the road, we see traffic signs in many shapes (triangle, square, octagon). In this chapter, we will explore some characteristics of geometry.

DEFINITIONS/VOCABULARY

Polygon: a many sided two-dimensional figure. In other words, it is a closed figure of three or more straight lines on a flat piece of paper. Examples are a triangle, a pentagon, and an octagon.

Edge: a side of the polygon. Edges is the plural of "edge." An edge is a line segment, that is, a small piece of a straight line.

Vertex: where two edges meet. Vertices is the plural of "vertex." A vertex forms an angle on the polygon.

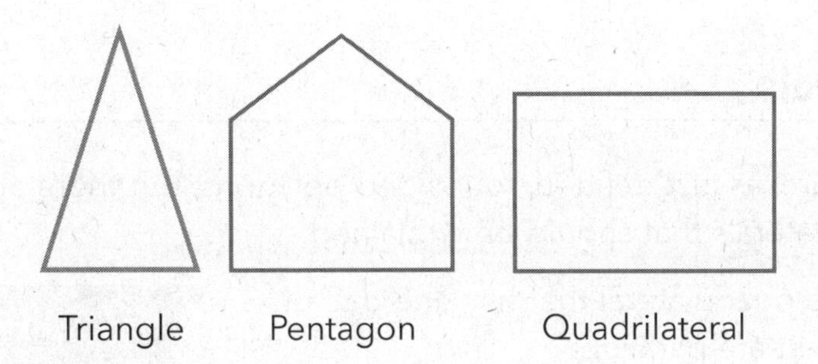

Polygons

There are many polygons. The following table names just a few and tells how many edges and vertices each one has.

Polygon	Edges	Vertices
Triangle	Three	Three
Quadrilateral	Four	Four
Pentagon	Five	Five
Hexagon	Six	Six
Octagon	Eight	Eight
Decagon	Ten	Ten

When we deal with polygons, the word "regular" is important. A **regular polygon** is a polygon that has all edges the same size, and the angles at each edge are the same size. A few examples of these are:

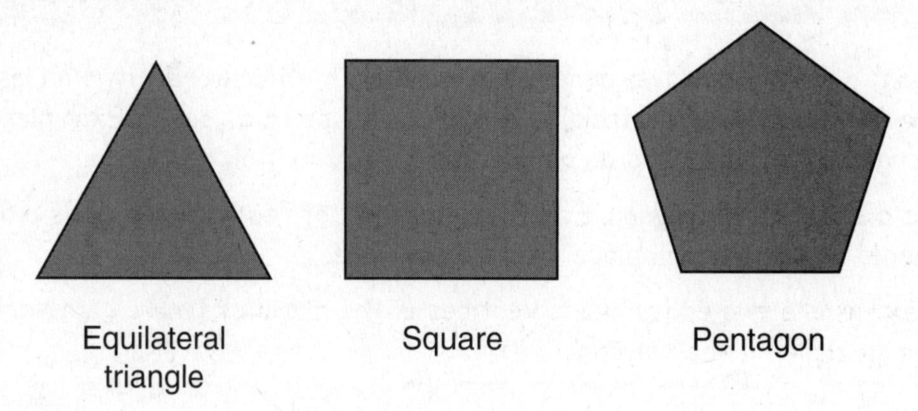

Quadrilaterals

Quadrilaterals are, as just defined, four-sided polygons, but there are five special types of quadrilaterals that should be explained.

A **trapezoid** is a quadrilateral that has one pair of opposite legs that are parallel.

A **parallelogram** is a quadrilateral that has both pairs of opposite legs parallel.

A **rhombus** is a quadrilateral that is a parallelogram, but in addition, all four legs are the same length.

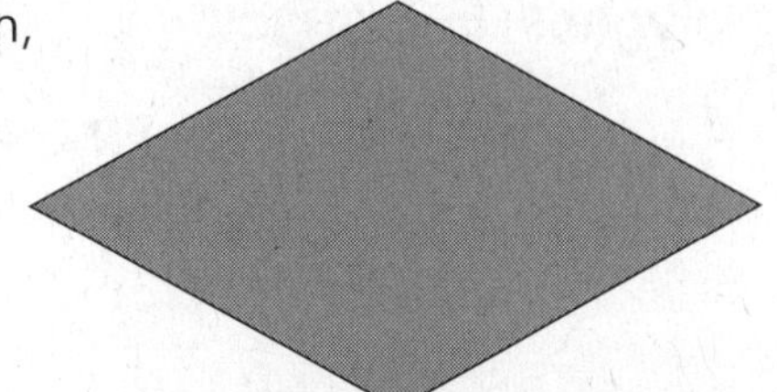

A **rectangle** is a quadrilateral that is a parallelogram, but in addition, all four angles are right angles (90°).

A **square** is a quadrilateral that combines one property of a rhombus with another of a rectangle. A square has all four sides that are equal in length, as well as four right angles.

Circles

Since they have no straight sides and no vertices, **circles** have special names for their parts. The edge of the circle is one continuous curved line that goes all the way around the circle. The **radius** of a circle is the distance from the center of the circle to the edge. The **diameter** of a circle is the distance from one edge of a circle, through the center, and over to the other edge. The diameter is, therefore, two times the size of the radius.

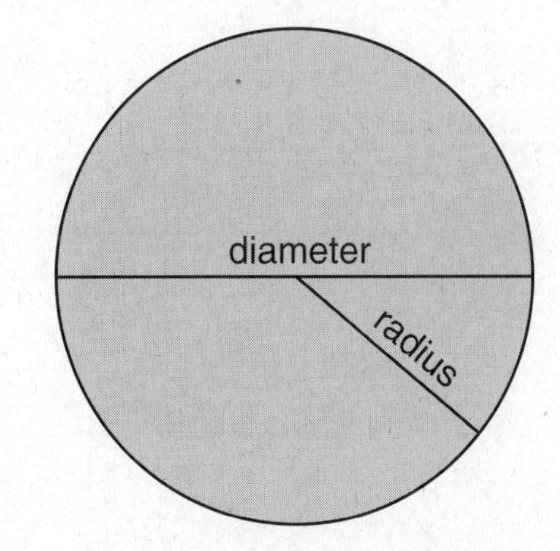

Polygons Exercises

1. Which of the following is a polygon?

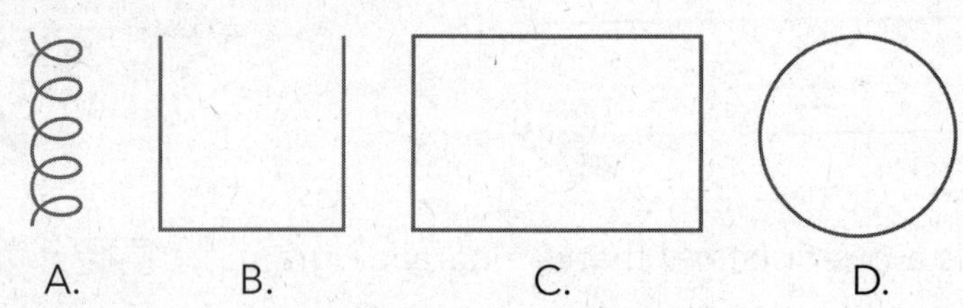

- ○ A.
- ○ B.
- ○ C.
- ○ D.

2. If you put two equilateral triangles together, what shape would you form?

- ○ A. A square
- ○ B. A rectangle
- ○ C. A hexagon
- ○ D. A rhombus

3. What polygon has six sides?

- ○ A. A hexagon
- ○ B. An octagon
- ○ C. A pentagon
- ○ D. A quadrilateral

4. Which shape is *not* a polygon? Circle your answer.

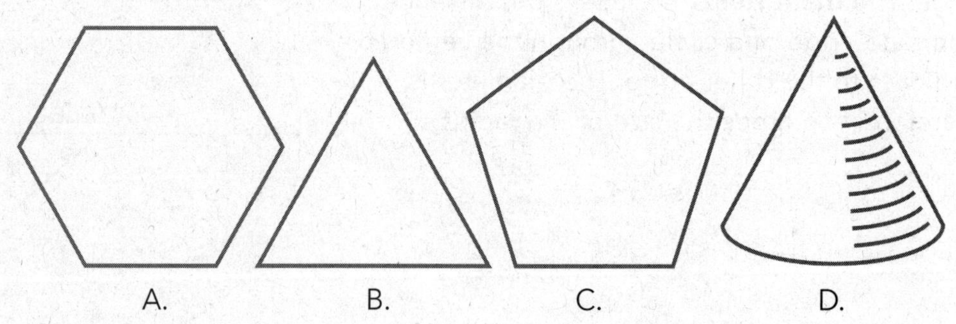

- ○ A.
- ○ B.
- ○ C.
- ○ D.

5. Tell how many sides each polygon has. Name each polygon and write it on the line.

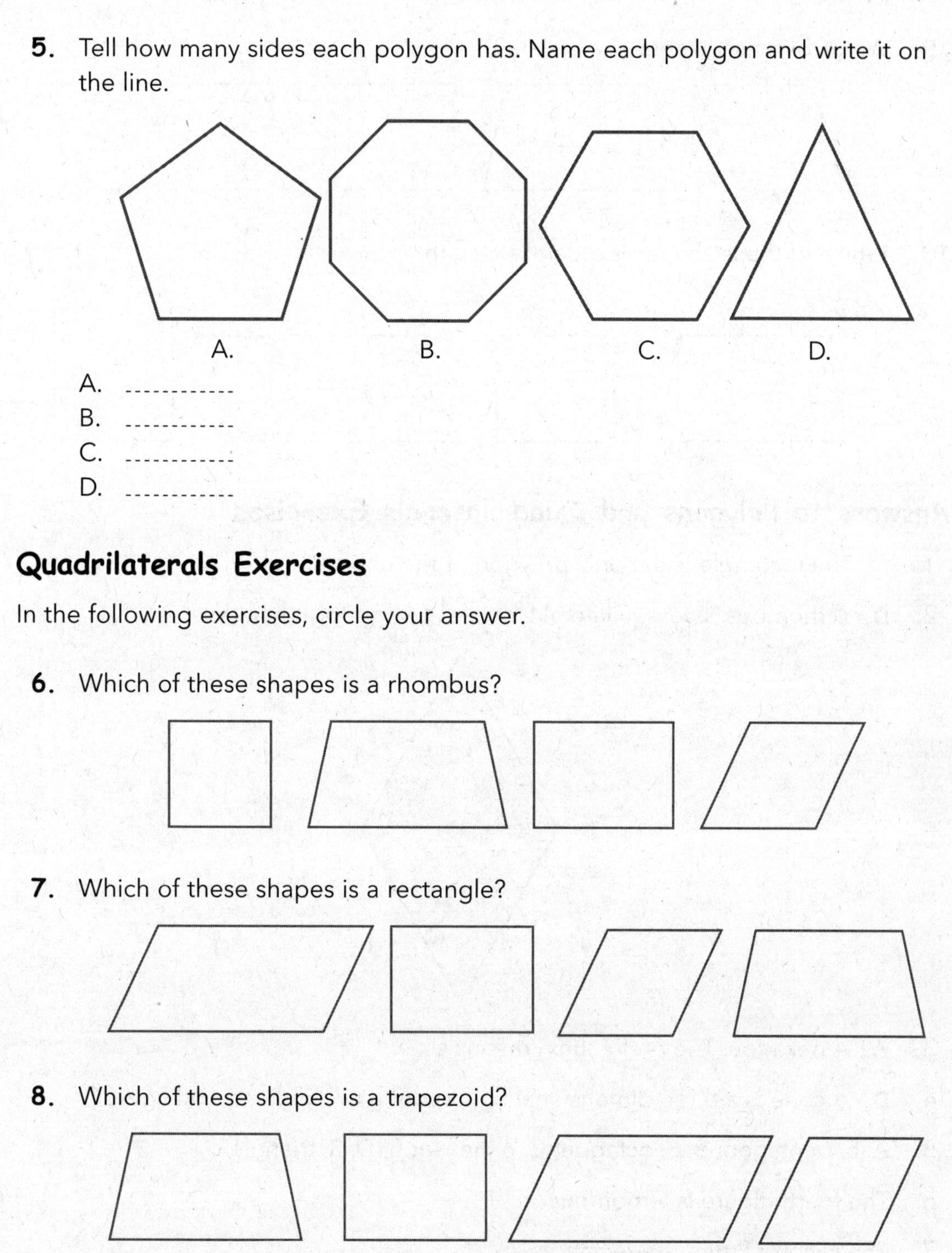

A. ------------
B. ------------
C. ------------
D. ------------

Quadrilaterals Exercises

In the following exercises, circle your answer.

6. Which of these shapes is a rhombus?

7. Which of these shapes is a rectangle?

8. Which of these shapes is a trapezoid?

9. Which of these shapes is a square?

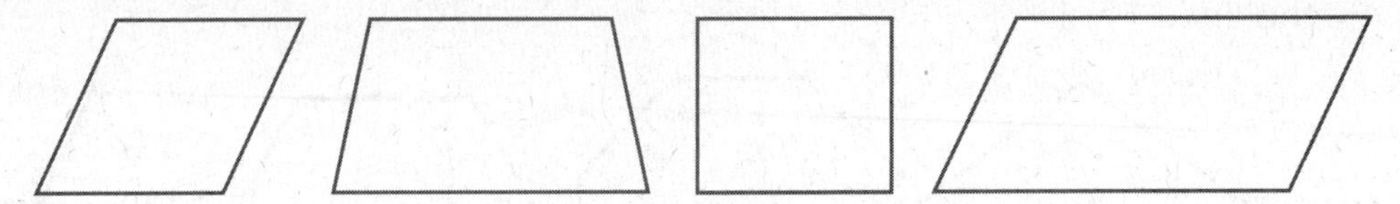

10. Which of these shapes is a parallelogram?

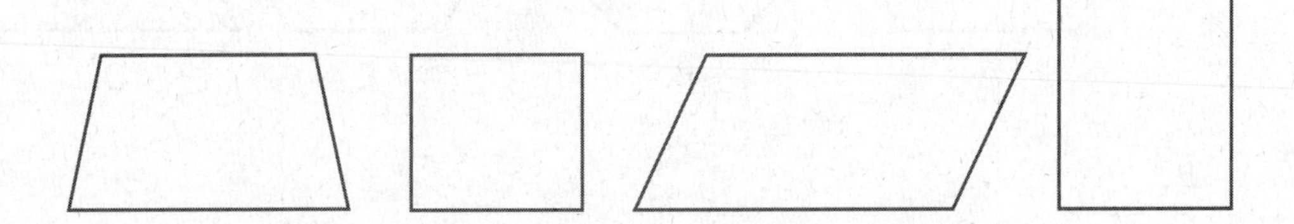

Answers to Polygons and Quadrilaterals Exercises

1. **C.** The rectangle is the only polygon of the choices.
2. **D.** A rhombus. Two equilateral triangles will go together, thus:

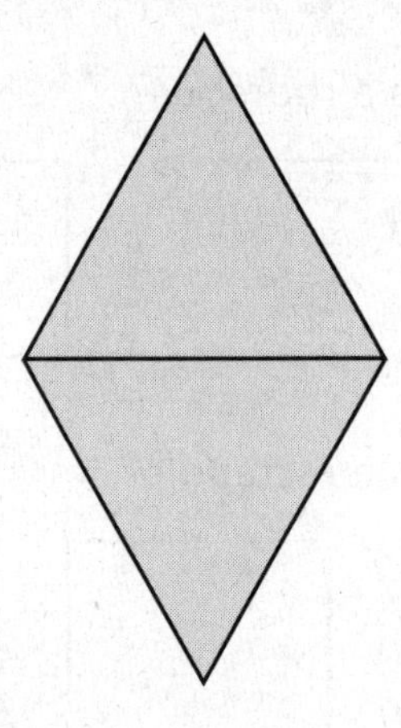

3. **A.** A hexagon. The prefix "hex" means six.
4. **D.** A cone is a three-dimensional figure, not a two-dimensional figure.
5. A. 5, pentagon; B. 8, octagon; C. 6, hexagon; D. 3, triangle.
6. The fourth figure is a rhombus.
7. The second figure is a rectangle.
8. The first figure is a trapezoid.
9. The third figure is a square.
10. The third figure is a parallelogram.

Perimeter of a Figure

The **perimeter** of a figure is the total distance around an object. Consider this figure:

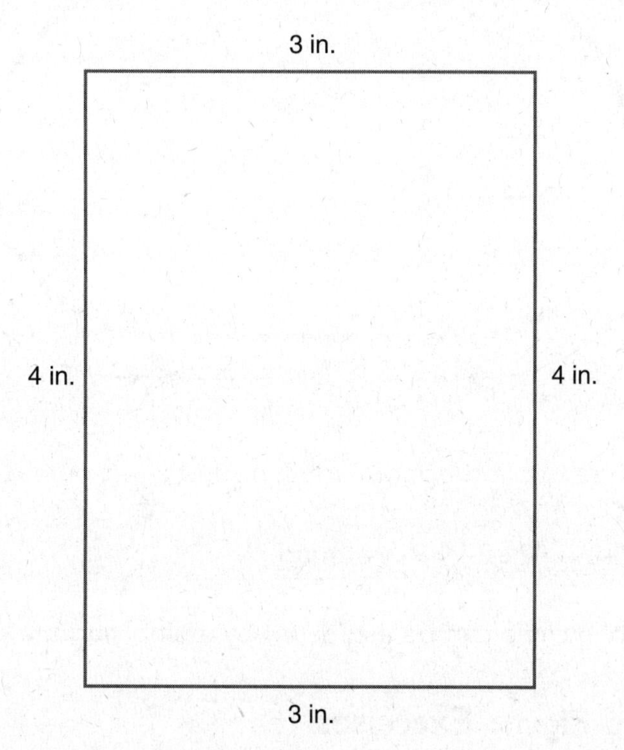

You see here that the perimeter of this rectangle is 3 + 3 + 4 + 4 = 14 inches around. Another way to look at this is to add one side to another and then double it. So (3 + 4) × 2 = 14 inches.

Let's look at another example:

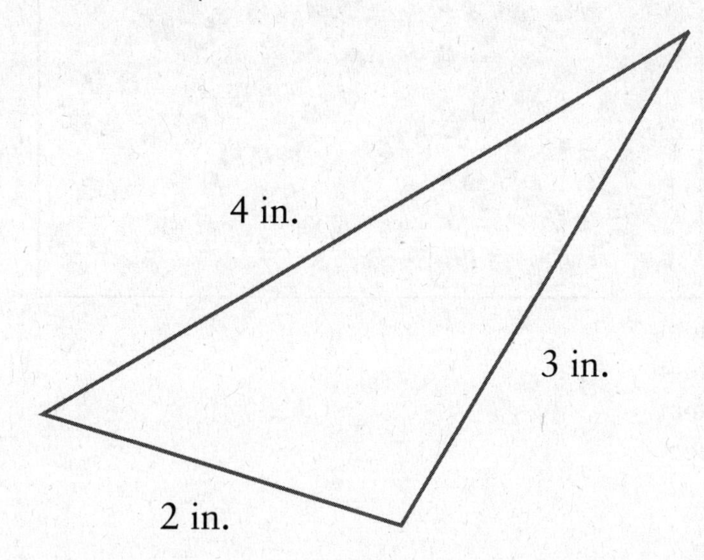

You see here that the perimeter of this triangle is 2 + 3 + 4 = 9 inches around.

Curved figures have a perimeter, as well.

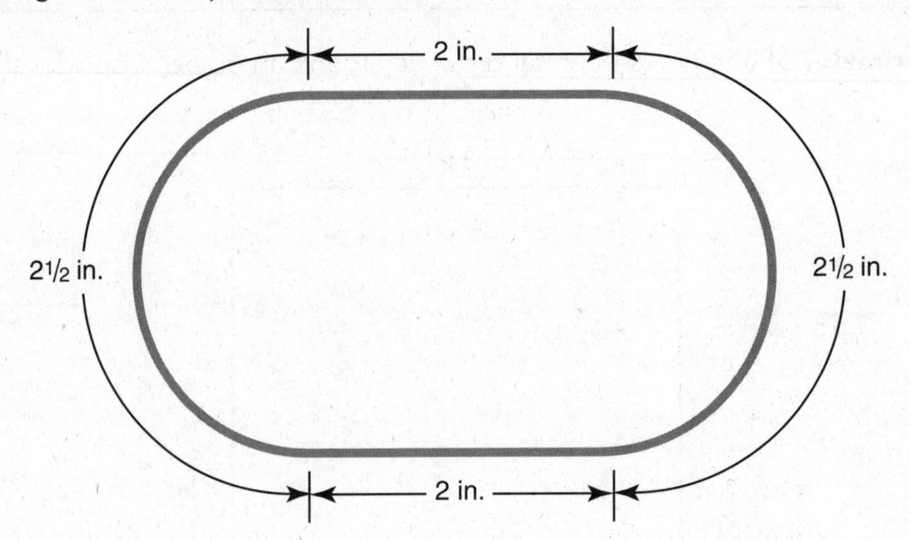

You see here that the top and bottom are 2 in., and $2\frac{1}{2}$ in. on each end, so that the perimeter is $2+2\frac{1}{2}+2+2\frac{1}{2}=9$ in. around.

The perimeter of a circle can be estimated by multiplying the diameter by 3.

Perimeter of a Figure Exercises

1. Sean is putting in a flower bed and wants to put a fence around it. The flower bed is shown below. How many feet of fence should Sean buy?

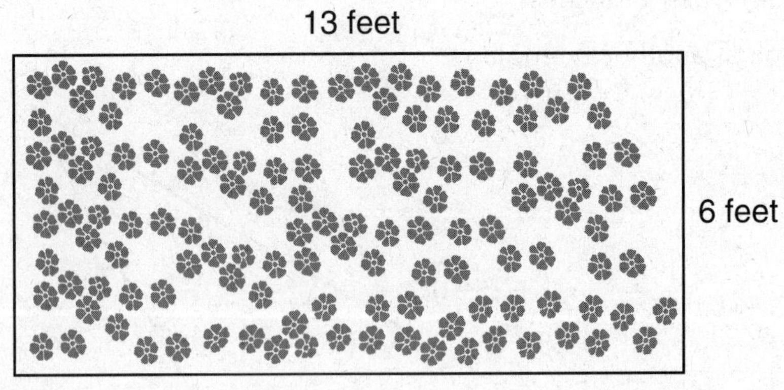

- A. 38 feet
- B. 32 feet
- C. 19 feet
- D. 78 feet

2. Gilda is decorating the top of a square box, shown below. She wants to put decorative paper around the box top. How much decorative paper does she need?

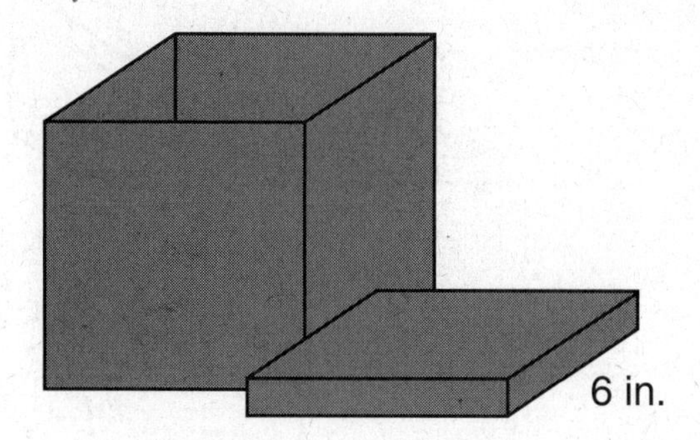

- ○ A. 36 inches
- ○ B. 30 inches
- ○ C. 28 inches
- ○ D. 24 inches

3. Leroy runs cross country. To practice, he runs around the school track, pictured below. How far around is this track?

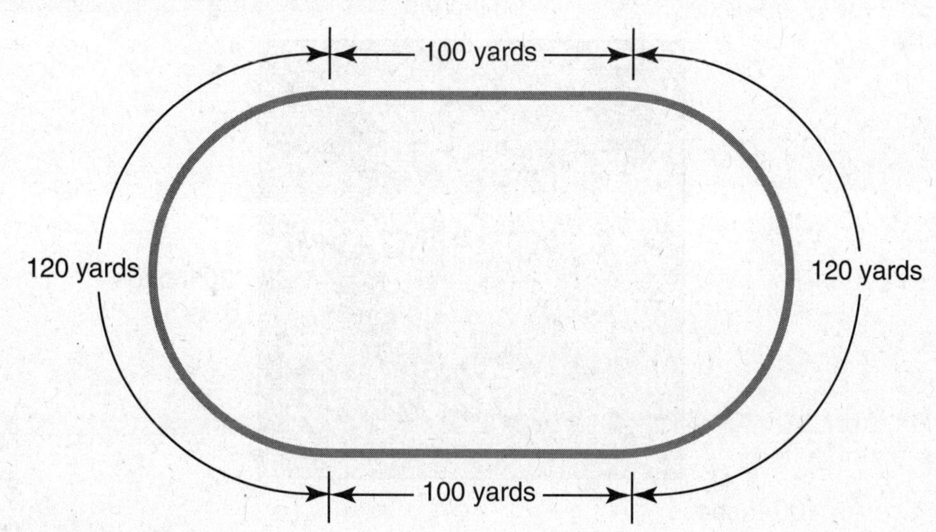

- ○ A. 220 yards
- ○ B. 340 yards
- ○ C. 440 yards
- ○ D. 400 yards

4. Andrea bought a pair of shoes. The rectangular box the shoes came in is pictured. What is the perimeter of the box?

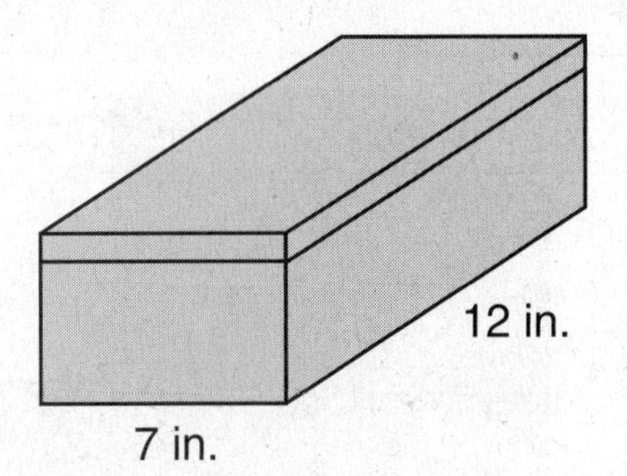

- ○ A. 40 inches
- ○ B. 38 inches
- ○ C. 34 inches
- ○ D. 19 inches

5. George is putting a frame around his favorite poster. The poster is shown below. How much framing material will George be getting?

- ○ A. 400 inches
- ○ B. 140 inches
- ○ C. 120 inches
- ○ D. 80 inches

6. Greg has odd-shaped tables in his restaurant. The tables are shaped like a trapezoid. What is the perimeter of the tables?

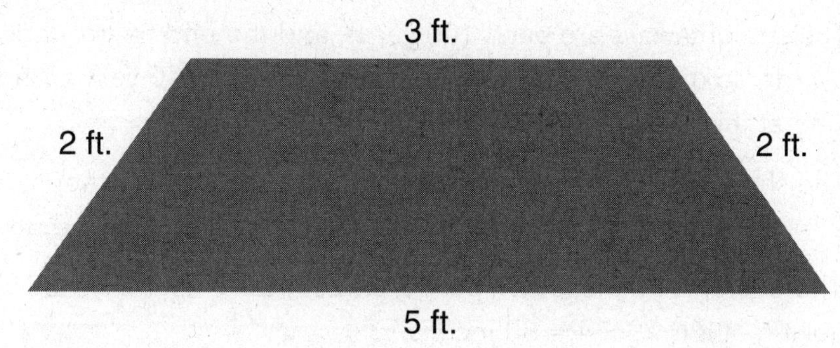

- ○ A. 6 feet
- ○ B. 10 feet
- ○ C. 12 feet
- ○ D. 18 feet

7. Joe sells stained glass windows. The one he has on display is a regular hexagon. What is the perimeter of the window?

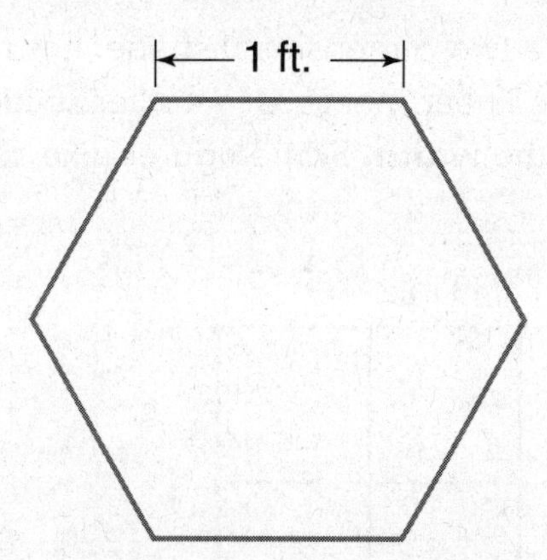

- ○ A. 8 feet
- ○ B. 6 feet
- ○ C. 4 feet
- ○ D. 10 feet

Answers to Perimeter of a Figure Exercises

1. **A.** The width of the flower bed is 6 feet, and the length of the bed is 13 feet. $6 + 13 + 6 + 13 = 38$ feet. Another way of looking at this: $(6 + 13) \times 2 = 38$ feet.

2. **D.** The box is 6 inches on a side. $6 + 6 + 6 + 6 = 24$ inches. Another way of looking at this is $6 \times 4 = 24$ inches.
3. **C.** The straightaways are each 100 yards, and the end half-rounds are each 120 yards, so the track is $100 + 120 + 100 + 120 = 440$ yards. Another way of looking at this is $(100 + 120) \times 2 = 440$ yards.
4. **B.** The shoebox is 7 inches on one side, 12 inches on another side. $7 + 12 + 7 + 12 = 38$ inches. Another way of looking at it is $(7 + 12) \times 2 = 38$ inches.
5. **D.** The poster is 20 inches on a side, so $20 + 20 + 20 + 20 = 80$. Another way of looking at it is $20 \times 4 = 80$ inches.
6. **C.** Adding the four lengths: $3 + 2 + 5 + 2 = 12$ feet. And so the perimeter is 12 feet.
7. **B.** Each side is 1 foot, so the perimeter is $1 + 1 + 1 + 1 + 1 + 1 = 6$ feet.

Area of Shapes on a Square Grid

Area is the measure of a two-dimensional shape. It is measured in square units. This is different from the linear measure in earlier sections. To measure area, simply multiply the **length** by the **width.** Some figures, like squares and rectangles, are easy to measure, thus:

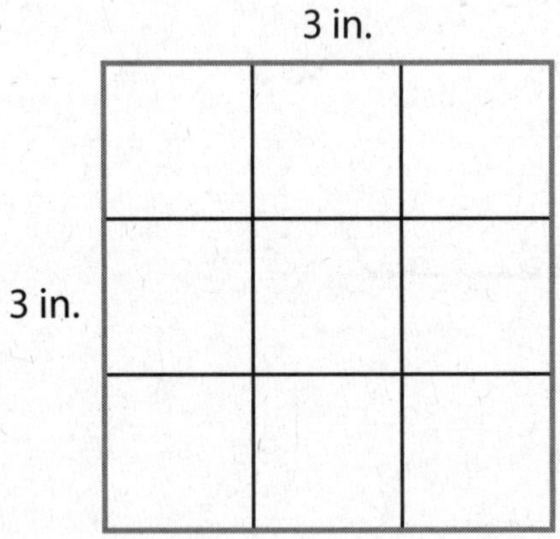

3 in.

3 in.

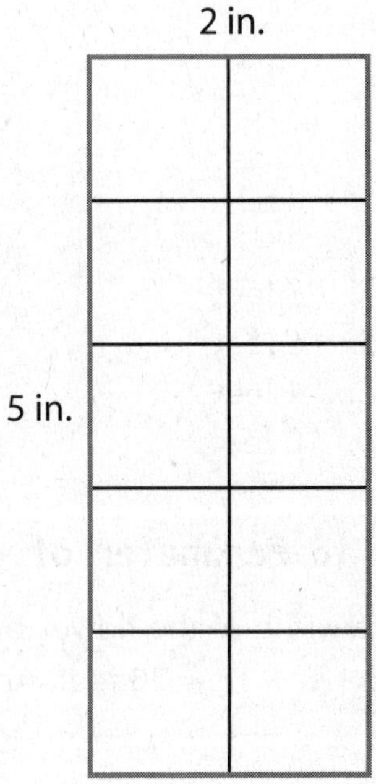

2 in.

5 in.

You see that the square is 3 inches on each side; you could count the squares inside to get 9. There is also an equation for the area of the square:

$$A = 3 \times 3 = 9 \text{ square inches, which is also written as } 9 \text{ in.}^2$$

In similar fashion, the rectangle is 2 inches on one side, 5 inches on the other side, so counting the squares gives you 10. Again, there is an equation for the area of the rectangle:

$$A = 2 \times 5 = 10 \text{ square inches, which is also written as } 10 \text{ in.}^2$$

You can see, therefore, that square measure is a *different kind* of measure than linear measure. Square measurements always use square units, whether it be square inches (in.^{2}), square feet (ft^2), or square centimeters (cm^2).

Area of Shapes on a Square Grid Exercises

> **Note: When doing these exercises, make sure you write down which units the problem is displaying. Always use the correct units that the exercises require.**

1. The CD case above has a grid laid over it. Each square is 1 square inch. Calculate the area of the CD in square inches.

Answer: ....

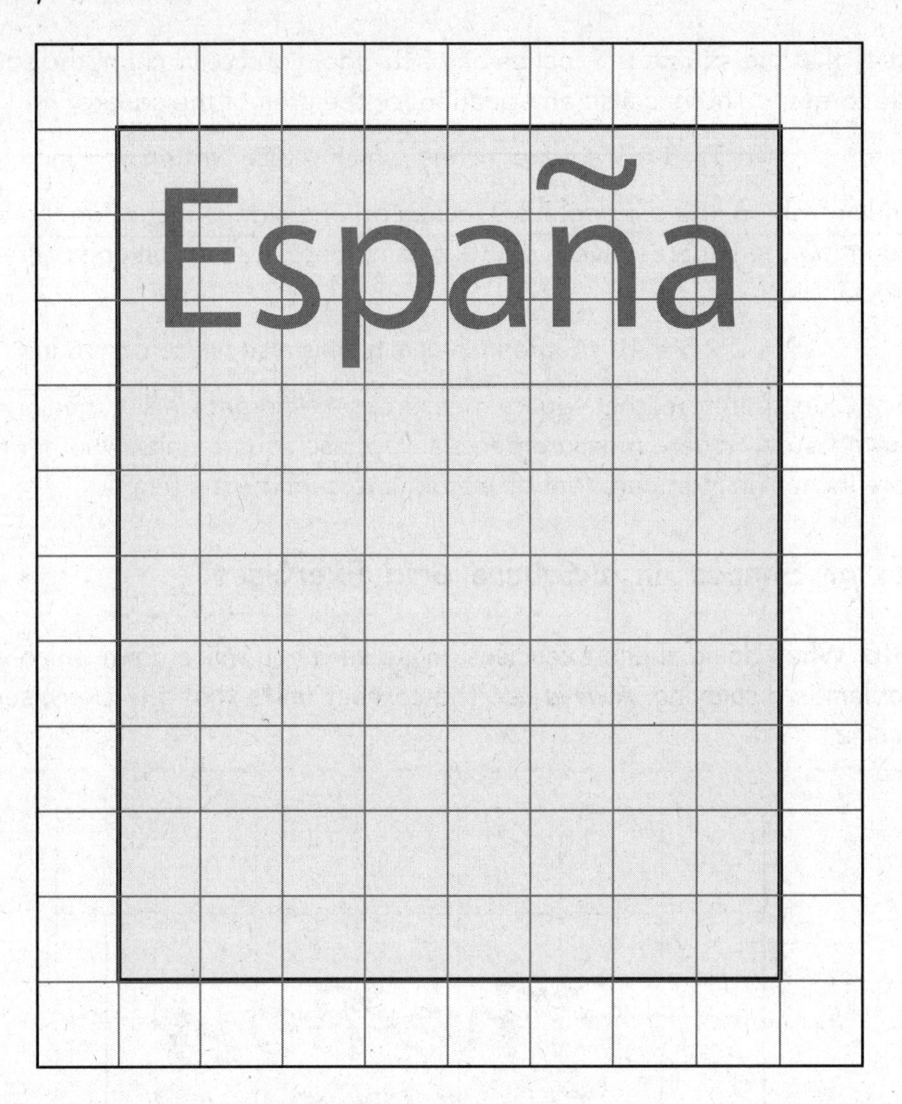

2. The book above has a grid laid over it. Each square is 1 square inch. Calculate the area of the book in square inches.

Answer: ____________________

3. The same book is ______ inches wide and ______ inches high. Multiply these two, and place your answer here ______. Is it the same as the area you calculated in question 2?

4. The cereal box above has a grid laid over it. Each square is 1 square inch. Calculate the area of the cereal box in square inches.

Answer: ....

5. The cereal box is ______ inches wide and ______ inches high. Multiply these two, and place your answer here ______. Is it the same area as you calculated in question 4?

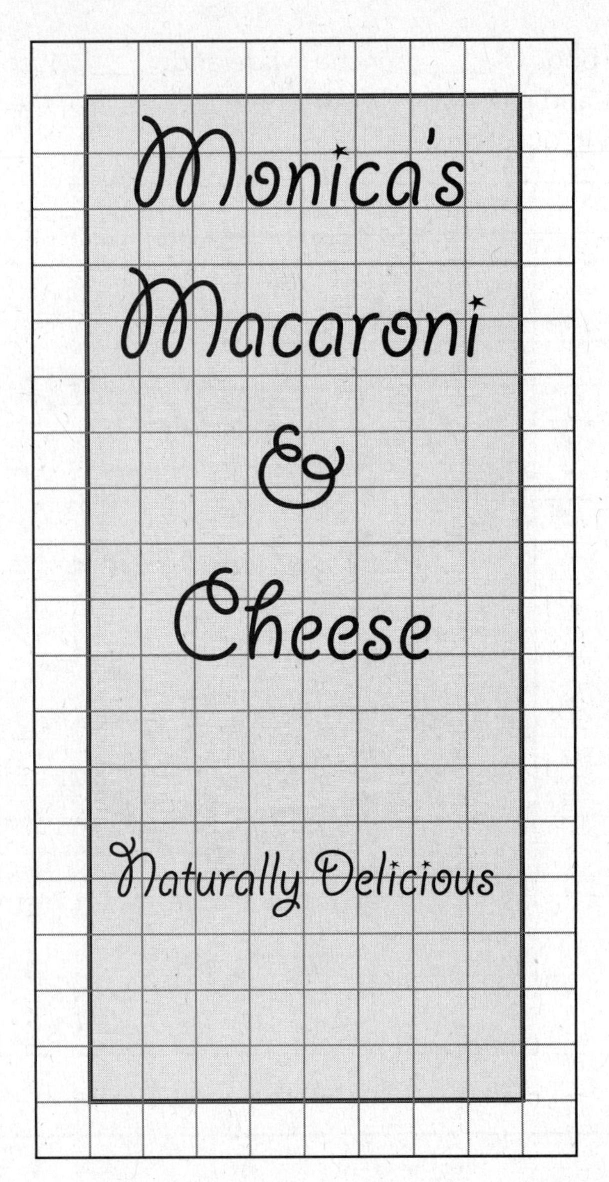

6. The macaroni and cheese box above has a grid laid over it. Each square is 1 square centimeter. Calculate the area of the box in square centimeters.

Answer: ____

7. The macaroni and cheese box is _______ cm wide and _______ cm high. Multiply these two, and place your answer here _______. Is it the same area as you calculated in question 6?

8. The flyer above has a grid laid over it. Each square is 1 square inch. Calculate the area of the flyer in square inches.

Answer: --

9. The flyer is ______ inches wide and ______ inches high. Multiply these two, and place your answer here ______. Is it the same area as you calculated in question 8?

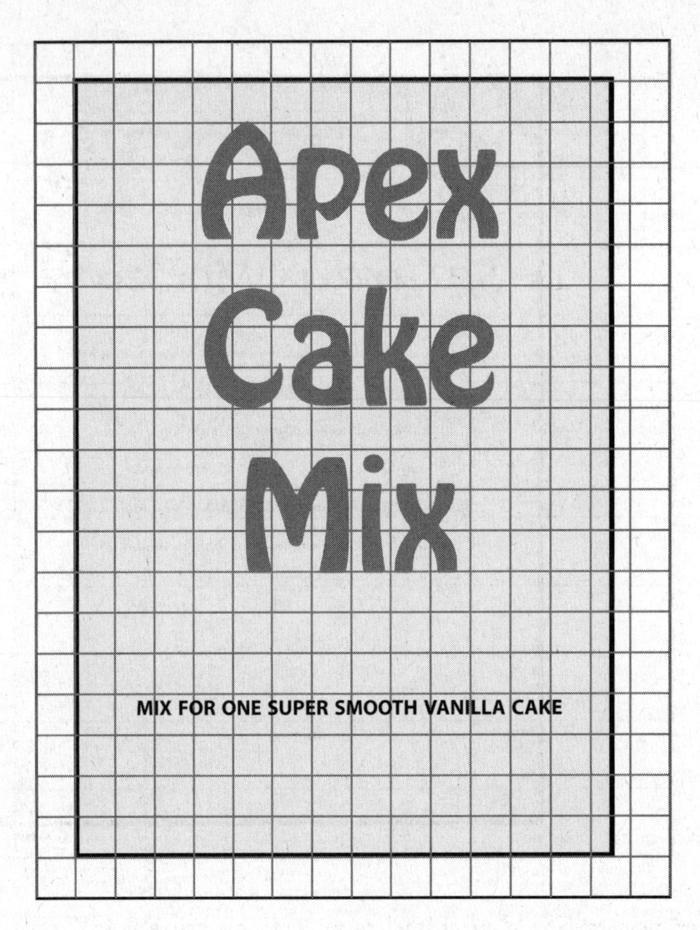

10. The cake box above has a grid laid over it. Each square is 1 square centimeter. Calculate the area of the cake box in square centimeters.

Answer: --

11. The cake box is _______ cm wide and _______ cm high. Multiply these two, and place your answer here _______. Is it the same area as you calculated in question 10?

12. The cracker box above has a grid laid over it. Each square is 1 square inch. Calculate the area in square inches of the cracker box.

Answer: --

13. The cracker box is _______ inches wide and _______ inches high. Multiply these two, and place your answer here _______. Is it the same area as you calculated in question 12?

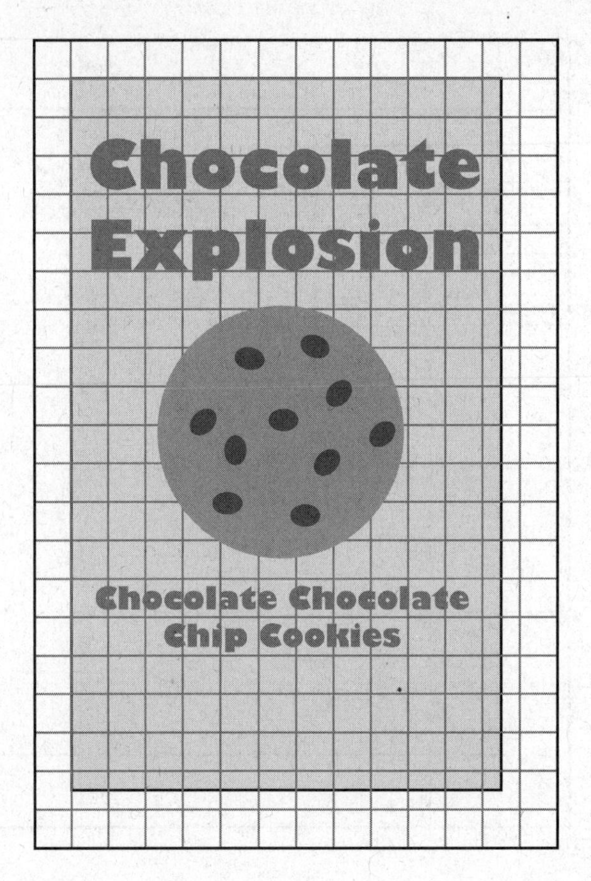

14. The cookie box above has a grid laid over it. Each square is 1 square centimeter. Calculate the area in square centimeters of the cookie box.

Answer: ______________________________

15. The cookie box is _______ cm wide and _______ cm high. Multiply these two, and place your answer here _______. Is it the same area as you calculated in question 14?

16. Lynn is having a picnic in the park. She spreads out her blanket as shown. What is the area it covers?

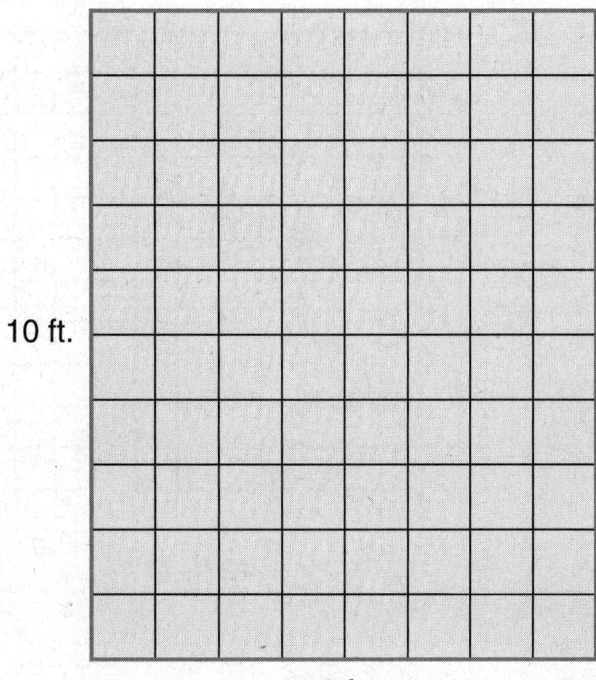

- ○ A. 18 ft^2
- ○ B. 30 ft^2
- ○ C. 60 ft^2
- ○ D. 80 ft^2

17. A wrestling mat is spread out in the gym for practice. What is the area of it?

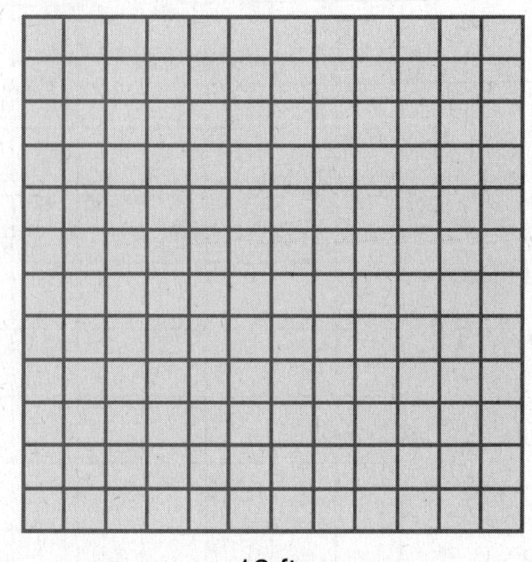

- A. 120 ft^2
- B. 144 ft^2
- C. 164 ft^2
- D. 174 ft^2

18. Hattie likes to display her jewelry collection on a felt cloth. What is its area?

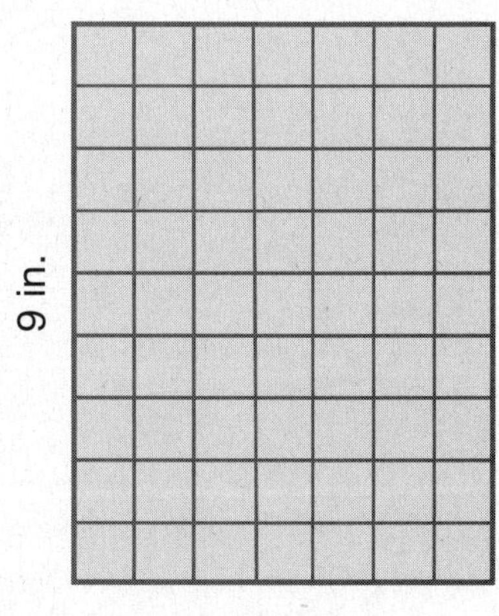

- A. 63 $in.^2$
- B. 60 $in.^2$
- C. 74 $in.^2$
- D. 70 $in.^2$

19. Which 2 of the following figures each has an area of 20 square inches (each square is one square inch)?

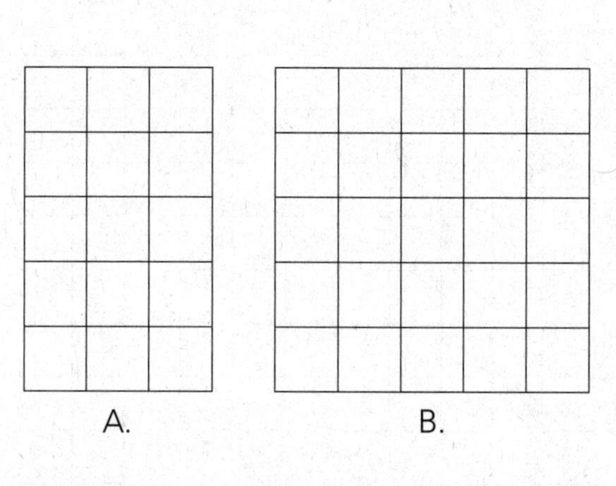

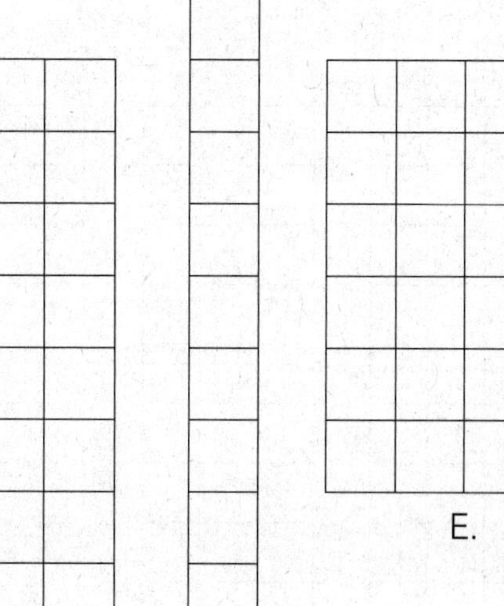

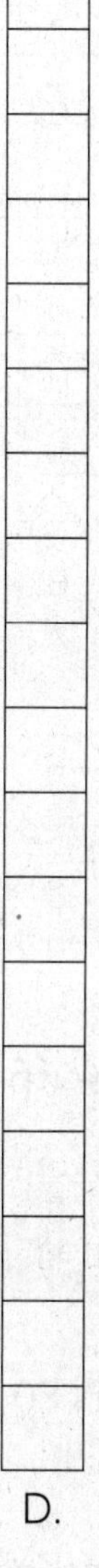

- ☐ A.
- ☐ B.
- ☐ C.
- ☐ D.
- ☐ E.

20. Which 2 of the following figures each has an area of 24 square inches (each square is one square inch)?

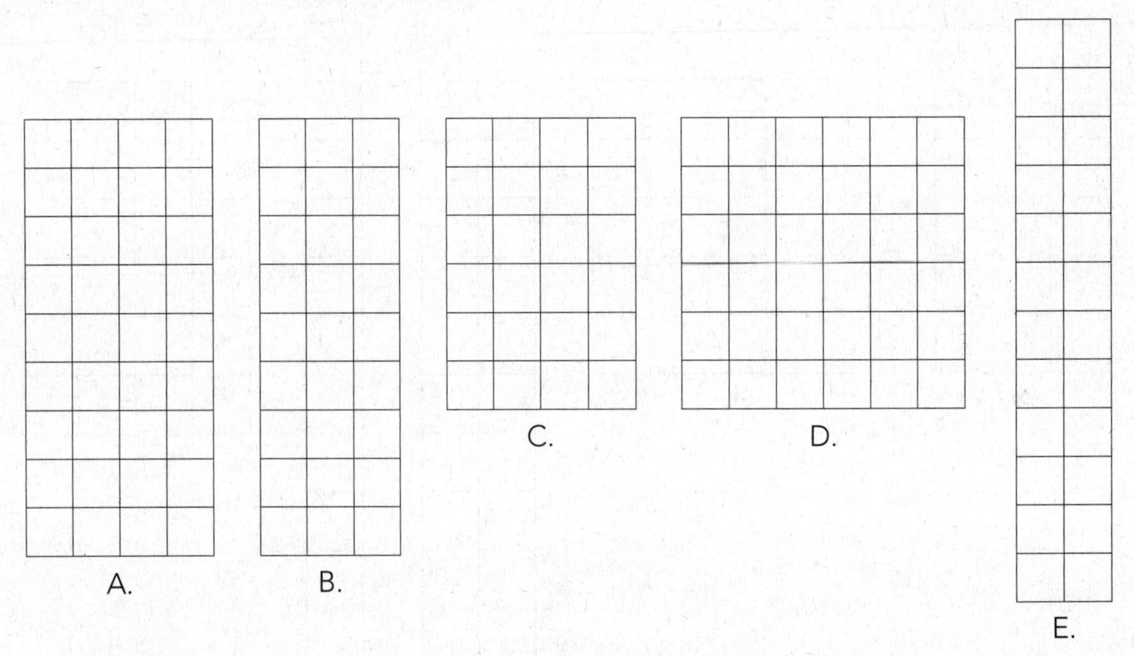

☐ A.
☐ B.
☐ C.
☐ D.
☐ E.

Answers to Area of Shapes on a Square Grid Exercises

1. The CD case covers 25 square inch blocks, and 5 one-half square inch blocks, so the area is: $25+2\frac{1}{2}=27\frac{1}{2}$ square inches, or $27\frac{1}{2}$ in.^{2}.

2. The book covers 80 square inch blocks, so the area of the book is 80 square inches, or 80 in.^{2}.

3. The book is 8 inches wide and 10 inches high. The book, therefore, is $8 \times 10 = 80$ in.^{2}. The answer is the same as in question 2.

4. The cereal box covers 70 square inch blocks, and 10 one-half inch square blocks, so the area is: $70 + 5 = 75$ square inches, or 75 in.^{2}.

5. The cereal box is $7\frac{1}{2}$ inches wide and 10 inches high. The cereal box, therefore, is $7\frac{1}{2} \times 10 = 75$ square inches, or 75 in.^{2}. The answer is the same as in question 4.

6. The macaroni and cheese box covers 144 square centimeter blocks, so the box is 144 square centimeters, or 144 cm^2.

7. The macaroni and cheese box is 8 centimeters wide and 18 centimeters high. The macaroni and cheese box, therefore, is $8 \times 18 = 144$ square centimeters, or 144 cm^2. The answer is the same as in question 6.

8. The flyer covers 80 square inch blocks, 26 one-half inch blocks, and 2 one-quarter inch blocks, so the flyer is $80 + 13 + \frac{1}{2}$ square inches, or $93\frac{1}{2}$ in.^{2}.

9. The flyer is $8\frac{1}{2}$ inches wide and 11 inches high. The flyer, therefore, is $8\frac{1}{2} \times 11 = 93\frac{1}{2}$ square inches, or $93\frac{1}{2}$ in.^{2}. The answer is the same as in question 8.

10. The cake box covers 247 square centimeter blocks, and 19 one-half square centimeter blocks, so the box is $247 + 9\frac{1}{2} = 256\frac{1}{2}$ square centimeters, or $256\frac{1}{2}$ cm^2.

11. The cake box is $13\frac{1}{2}$ centimeters wide and 19 centimeters high. The cake box, therefore, is $13\frac{1}{2} \times 19 = 256\frac{1}{2}$ square centimeters, or $256\frac{1}{2}$ cm^2. The answer is the same as in question 10.

12. The cracker box covers 63 square inch blocks, and 9 one-half inch blocks. Therefore, the area is $63 + 4\frac{1}{2} = 67\frac{1}{2}$ square inches, or $67\frac{1}{2}$ in.^{2}.

13. The cracker box is $7\frac{1}{2}$ inches wide and 9 inches high. The cracker box, therefore, is $7\frac{1}{2} \times 9 = 67\frac{1}{2}$ square inches, or $67\frac{1}{2}$ in.^{2}. The answer is the same as in question 12.

14. The cookie box covers 198 square centimeter blocks, 29 one-half square centimeter blocks, and 1 one-quarter centimeter block, so the box is $198 + 14\frac{1}{2} + \frac{1}{4} = 212\frac{3}{4}$ square centimeters, or $212\frac{3}{4}$ cm^2.

15. The cookie box is $11\frac{1}{2}$ centimeters wide and $18\frac{1}{2}$ centimeters high. The cookie box, therefore, is $11\frac{1}{2} \times 18\frac{1}{2} = 212\frac{3}{4}$ square centimeters, or $212\frac{3}{4}$ cm^2. The answer is the same as in question 14.

16. **D.** 80 ft^2.

17. **B.** 144 ft^2.

18. **A.** 63 in.^{2}.

19. **C** and **D**.

20. **C** and **E**.

Make Multiplication Easier by Using the Distributive Property

The distributive property can help make multiplication problems easier to calculate. Let's look at an example.

Example: Jessica is tiling a wall in her kitchen. She has measured the wall, and it will be able to take 8 tiles in a row, and 7 rows, but she does not know what 8×7 is. How many tiles will she need to complete the job?

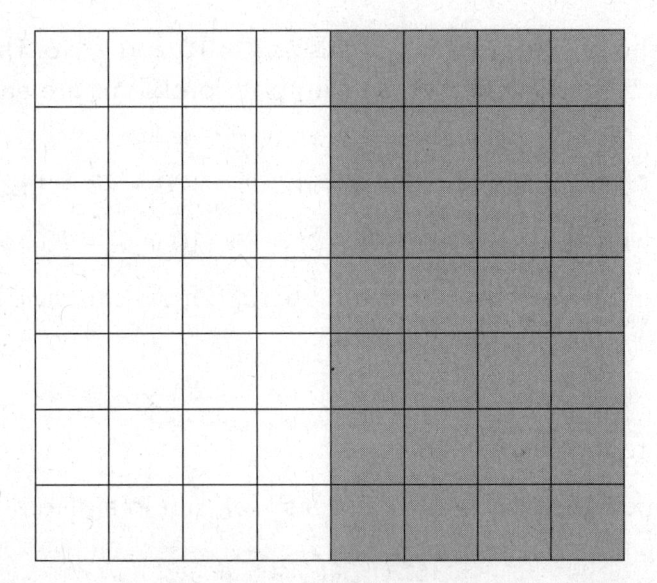

What Jessica can do is to break the problem up into two multiplication problems, as shown above. She sees that the first 4 columns, with 7 tiles in each column, is 28. Then, the last 4 columns, with 7 in each column, is also 28. Then 28 + 28 = 56. This is an illustration of the distributive property:

$$(7 \times 4) + (7 \times 4) = 7 \times (4 + 4) = 7 \times 8 = 56$$

Let's look at another example.

Example: The Kearney Kings Marching Band are in this year's July 4th Parade. There are 7 band players in a row, and there are 16 rows. How many band members are marching in this year's parade?

To solve this, we can break up the problem into two problems. Look at the picture (we'll use circles to show band members):

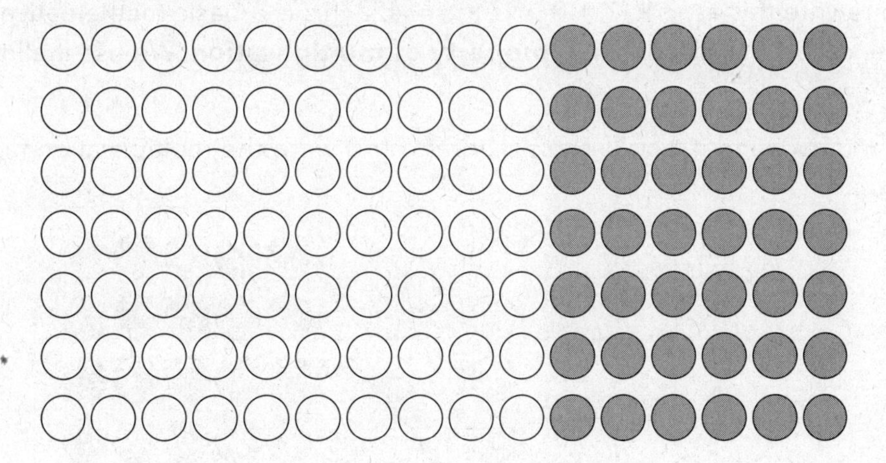

As the diagram shows, we can think of this as 7×10 and 7×6. Then we can add the two products to get the answer. Solving both problems are easier than solving the original one:

$$7 \times 10 = 70 \qquad 7 \times 6 = 42 \qquad 70 + 42 = 112$$

This can also be written as $(7 \times 10) + (7 \times 6) = 7 \times (10 + 6) = 70 + 42 = 112$

Now, some might think that 7×6 is a bit hard, but we can divide that problem further to

$$7 \times 3 = 21 \text{ and } 7 \times 3 = 21 \qquad 21 + 21 = 42$$

and we are back to the same 7×6.

Consider the multiplication fact 7×6. Let's look at it in a diagram:

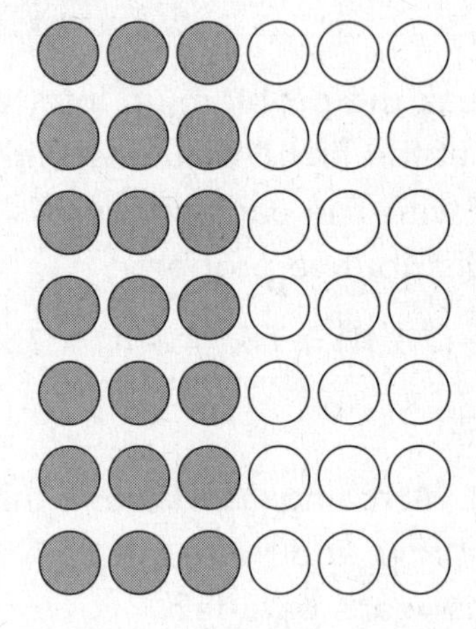

What we have done is to take the second multiplication fact in our last example, and separate it out. Look again, and see that $7 \times 6 = 7 \times 3 + 7 \times 3 = 42$.

You can rewrite this as: $7 \times (3 + 3) = 7 \times 6 = 42$. This is a basic mathematical principle, called the **distributive property of multiplication**. We use it all the time to make math calculations easier.

Now you try using it. Look at this diagram, and write the multiplication fact that it shows:

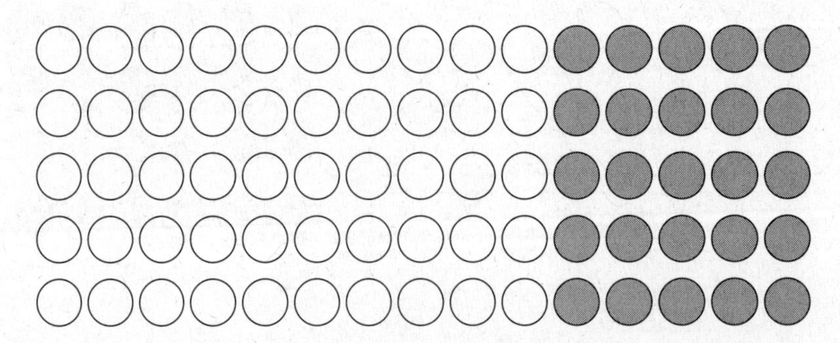

Write the fact here: ____ × ____ + ____ × ____ = ____ × (____ + ____)

Did you write $5 \times 10 + 5 \times 5 = 5 \times (10 + 5)$? This would describe the fact correctly.

Distributive Property Exercises

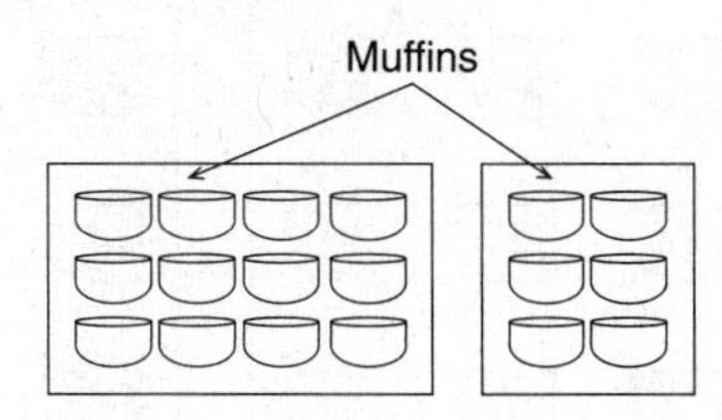

1. Grizelle baked some yummy cranberry muffins. Explain how these two trays Illustrate the multiplication fact 3×6.

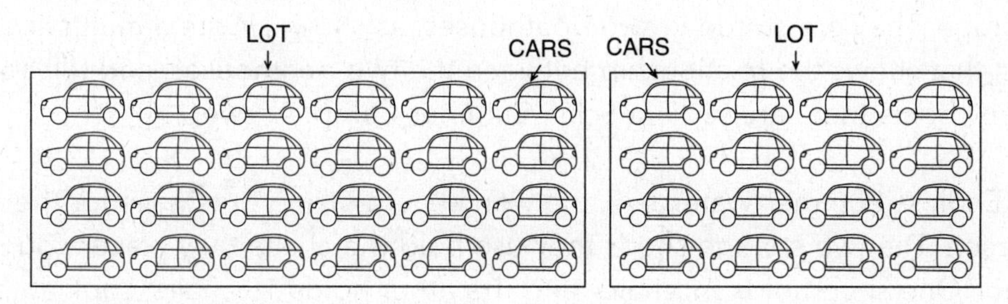

Explain your answer: ____________________

2. The parking lots pictured above suggest a number sentence. Which of these would it suggest? Explain your answer.

- ○ A. $6 + 4 + 4 + 4 =$ ___
- ○ B. $6 \times 4 \times 4 + 4 =$ ___
- ○ C. $4 \times 6 + 4 \times 4 =$ ___
- ○ D. $4 \times 6 + 4 + 6 =$ ___

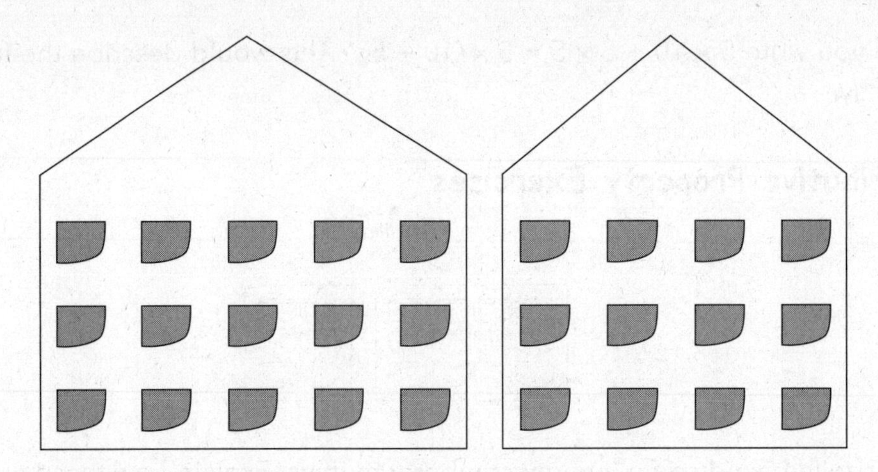

3. Up at the Schiff Summer Camp, the boats are being taken out of winter storage. They are stored in two boathouses, as shown. Write a multiplication fact that shows the relationship between the two boathouses, and tell how many boats there are in all at Schiff Summer Camp. Answer: ____

4. At Endless Summer Waterpark, the workers are taking out chairs for the season. The two storage sheds they use hold the chairs they place around the pool. One shed holds 21 chairs, and the other holds 18 chairs. Look at the following diagram, and tell whether it seems to represent this arrangement. If not, explain why not.

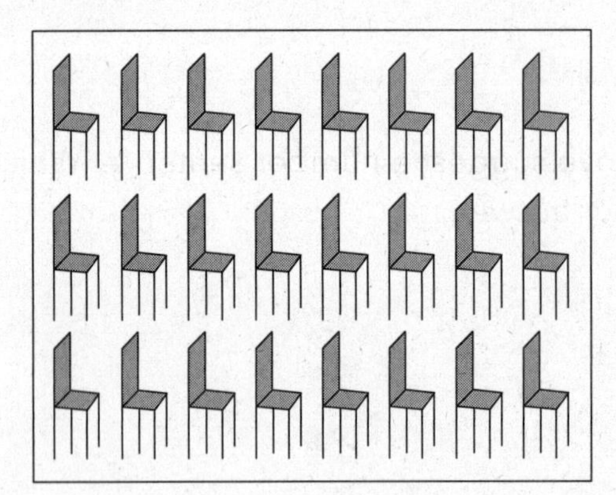

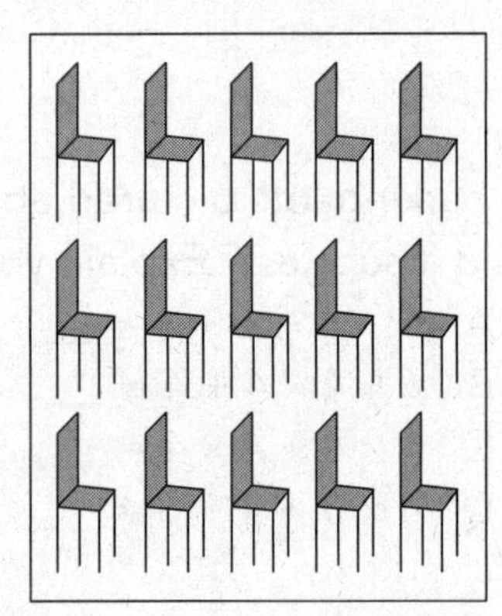

Explain your answer: --

5. At Glen Gray Scout Camp, Drew is making up trays of oranges for the lunch meal in the dining hall. He has two trays of oranges, one with 20 oranges in it, and one with 25 oranges in it. Draw a diagram showing how the oranges could be arranged, having both of the trays with the same number of rows in each.

Answers to Distributive Property Exercises

1. If we multiply 3×4, we'll get 12, and if we multiply 3×2, we'll get 6. Then, $(3 \times 4) + (3 \times 2) = 12 + 6 = 18$. This is an illustration of the distributive property.
2. **C.** $4 \times 6 + 4 \times 4 = 24 + 16 = 40$. The first parking lot has 4 rows of 6 cars each, and the second lot has 4 rows of 4 cars each. The first lot, then, has $4 \times 6 = 24$ cars in it, and the second lot has $4 \times 4 = 16$ cars in it. This can be rewritten as $4 \times (6 + 4)$. This is another illustration of the distributive property.
3. $(3 \times 5) + (3 \times 4)$. This can be written as $3 \times (5 + 4)$ so there are $3 \times 9 = 27$ boats, or, to put it another way, $3 \times 4 + 3 \times 5 = 12 + 15 = 27$.
4. It does *not* explain the diagram correctly. The problem describes 21 chairs in one shed, and 18 chairs in the second shed. That would indicate 3 rows of 7 chairs each, and 3 rows of 6 chairs each. The multiplication diagram shows 3 rows of 8 chairs, and 3 rows of 5 chairs. Though the sum is the same (39), the diagram does not show the chairs arranged as the problem is written.
5. 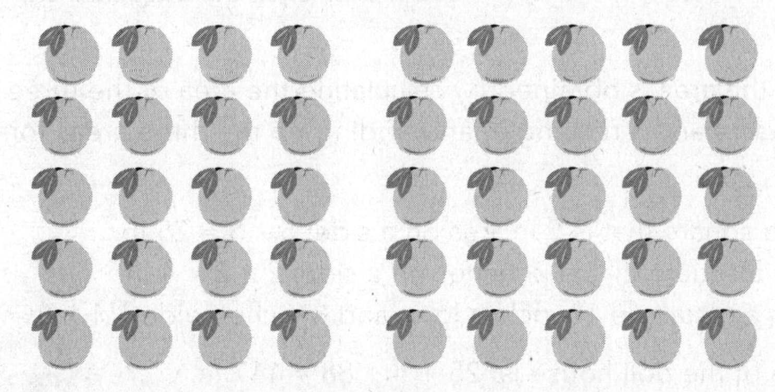

   This diagram shows how the oranges are arranged:

$$(5 \times 4) + (5 \times 5) = 20 + 25 = 45, \text{ or,}$$

$$(5 \times 4) + (5 \times 5) = 5 \times (4 + 5) = 5 \times 9 = 45$$

Area of Odd Shapes on a Square Grid

Some shapes are a combination of two or more rectangles or squares. When this happens, the best way to calculate the area is to break the shape into separate rectangles or squares. Then, the sum of the individual shapes can be added to obtain the total area of the shape.

For example, here is the floor area of a doll house:

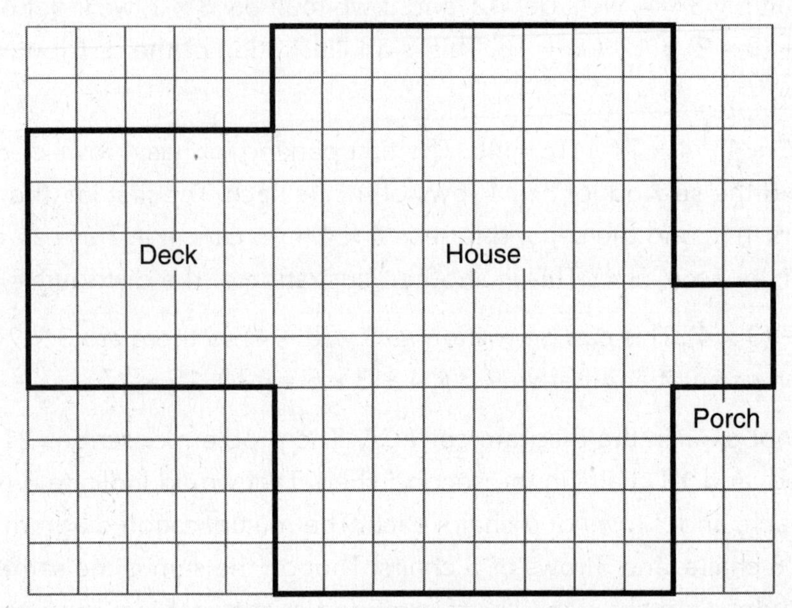

In this example, the area is obtained by calculating the area of the three separate shapes (two squares and a rectangle) and adding up the three areas for the total area.

The deck is a square that is 5 inches on a side: $5 \times 5 = 25$ in.^{2}
The porch is a square that is 2 inches on a side: $2 \times 2 = 4$ in.^{2}
The house is a rectangle 11 inches long and 8 inches wide: $11 \times 8 = 88$ in.^{2}

The total area of the doll house is: $25 + 4 + 88 = 117$ in.^{2}

Area of Odd Shapes on a Square Grid Exercises

1. Courtney has a house that has a patio out back, and a side entrance with a porch on it, as diagrammed below:

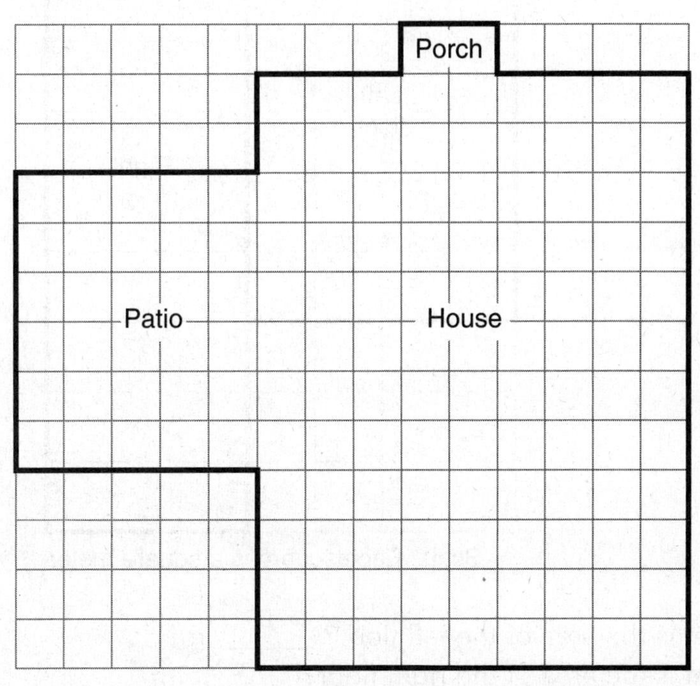

A. What is the area of the house? _______ yd^2
B. What is the area of the patio? _______ yd^2
C. What is the area of the porch? _______ yd^2
D. What is the area of the whole house? _______ yd^2

2. Bob was tiling the floor of two rooms in a house. These rooms were both rectangles as the following figure illustrates:

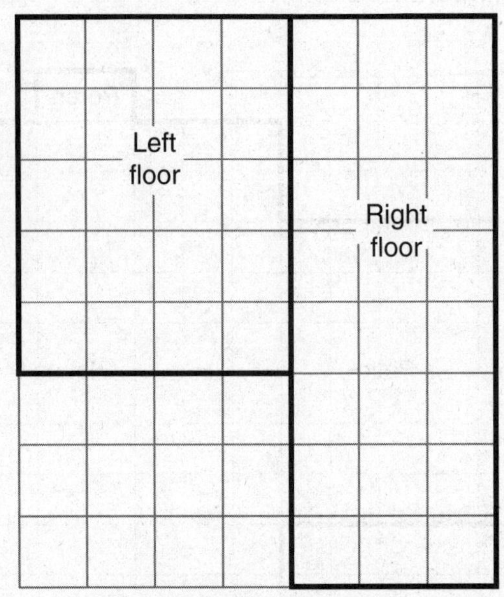

Scale: Each square is a square meter

A. What is the area of the left floor? _______ m^2

B. What is the area of the right floor? _______ m^2

C. What is the area of the two floors combined? _______ m^2

3. Andrew was mowing a lawn and noticed that the front yard he was mowing was a square, and the side yard he was mowing was a rectangle, as illustrated in the following diagram:

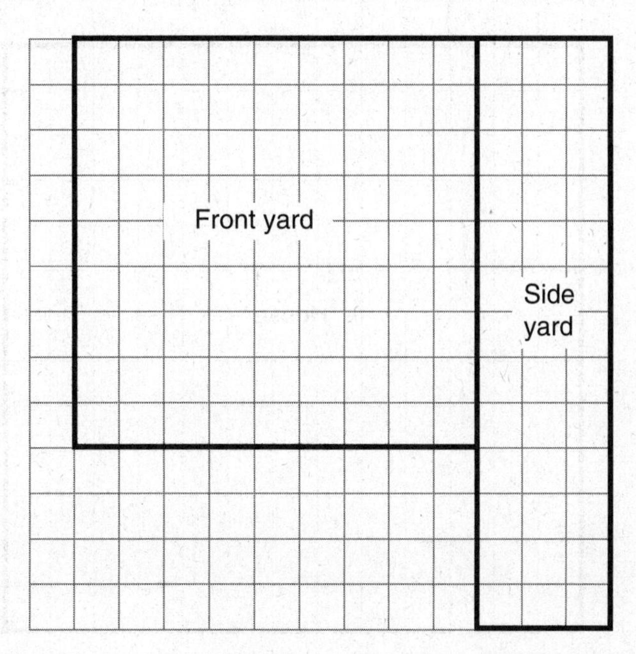

Scale: Each square is a square yard

A. What is the area of the front yard? _______ yd^2
B. What is the area of the side yard? _______ yd^2
C. What is the area of the two yards combined? _______ yd^2

4. Victorian houses often have wraparound porches. Here is an example:

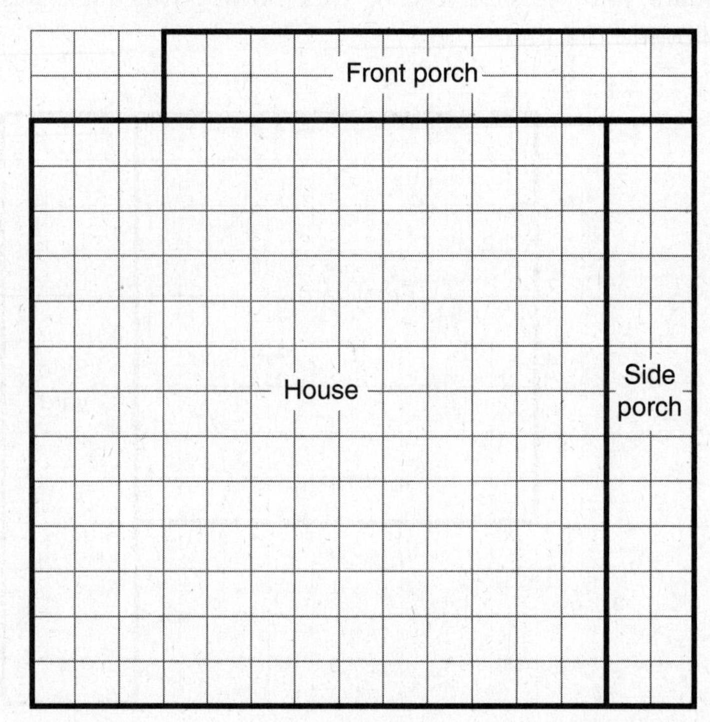

Scale: Each square is a square yard

A. What is the area of the house? ______ yd^2
B. What is the area of the front porch? ______ yd^2
C. What is the area of the side porch? ______ yd^2
D. What is the area of the house and porches combined? ______ yd^2

5. Angela was playing with blocks, and made a shape that combined three rectangles as shown in the following diagram:

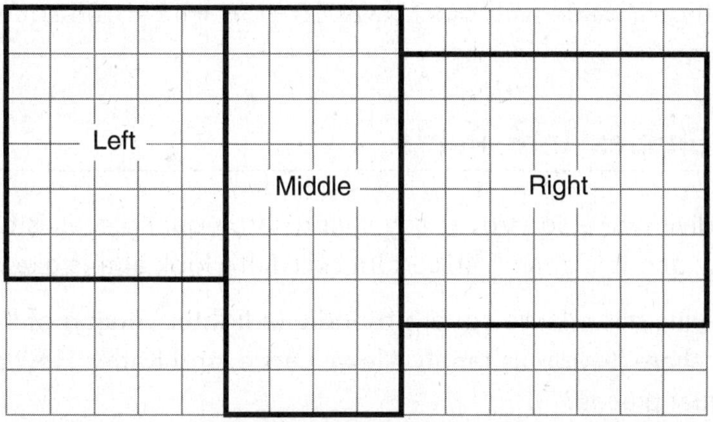

Left
Middle
Right

Scale: Each square is a square centimeter

A. What is the area of the left rectangle? _______ cm^2
B. What is the area of the middle rectangle? _______ cm^2
C. What is the area of the right rectangle? _______ cm^2
D. What is the area of the rectangles combined? _______ cm^2

Answers to Area of Odd Shapes on a Square Grid Exercises

1. Courtney's house covers 108 square yards. The patio covers 30 square yards. The porch covers 2 square yards. Therefore, the area of Courtney's house is 108 + 30 + 2 = 140 square yards, or 140 yd^2.

2. The floor Bob is tiling has two floors, a left and a right floor. The left floor covers 20 square meters. The right floor covers 24 square meters. Therefore, the area of the two floors combined is 20 + 24 = 44. The floor is 44 square meters, or 44 m^2.

3. The lawn Andrew is mowing is two areas. The front yard covers 81 square yards, and the side yard covers 39 square yards. Therefore, the area of the two yards combined is 81 + 39 = 120 square yards, or 120 yd^2.

4. The Victorian house covers 169 square yards. The front porch covers 24 square yards. The side porch covers 26 square yards. Therefore, the area of the Victorian house is 169 + 24 + 26 = 219 square yards, or 219 yd^2.

5. The shape Angela made formed three rectangles. The left rectangle covered 30 square centimeters. The middle rectangle covered 36 square centimeters. The right rectangle covered 42 square centimeters. Therefore, the area of the shape Angela made with blocks was 30 + 36 + 42 = 108 square centimeters, or 108 cm^2.

Dividing Shapes into Parts

Cutting (or dividing) a square, or any shape, into equal pieces is called **partitioning**, and it is a most important skill. Let's look at this example.

Example: Willie has a large square blondie (a lighter version of the brownie), and he wants to share it with his family Alexa, Lance, and Kathy. How can he divide it into four equal pieces?

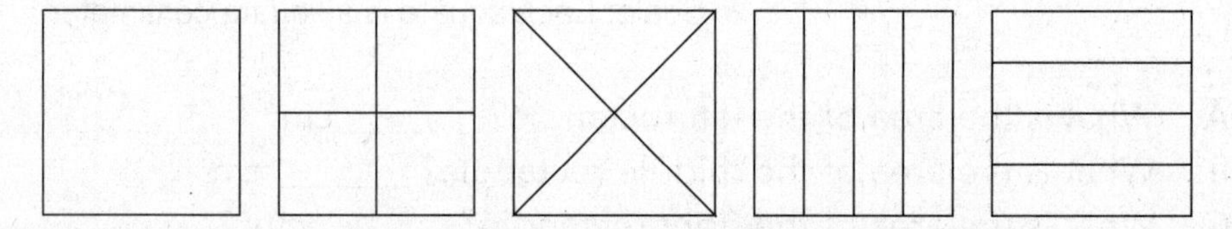

Here are four ways of dividing the blondie. Now, Willie probably would not want to cut in in strips as in the last two examples, but he could. Here's another way of cutting it into four equal pieces:

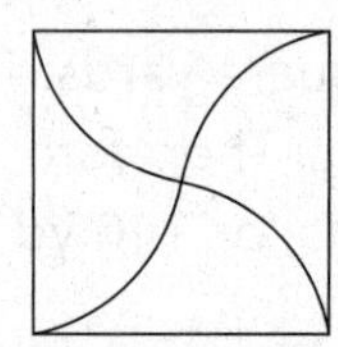

Of course, this is difficult to cut, but it indicates that the four ways the blondie was cut in the first figure are not the only ways of cutting it. There are, in fact, many ways of cutting the blondie, but we are just considering straight cuts.

Let's look at another shape: a rectangle:

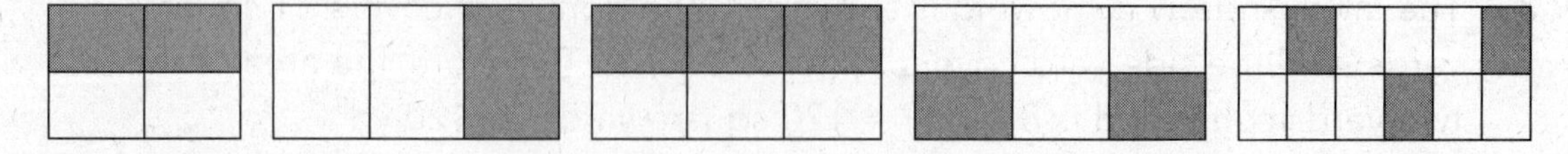

In the above diagram, the five rectangles are partitioned into 4 pieces, 3 pieces, 6 pieces, and 12 pieces. As shown, the rectangles are shaded:

The first rectangle is 2/4 shaded.
The second rectangle is 1/3 shaded.
The third rectangle is 3/6, or 1/2, shaded.

The fourth rectangle is 2/6, or 1/3, shaded.
Finally, the fifth rectangle is 3/12, or 1/4, shaded.

The rectangles, which have sections shaded, revisit an important idea: The same number can be named in many different ways. There are, in fact, an infinite number of ways of naming a number.

Circles, too, can be partitioned:

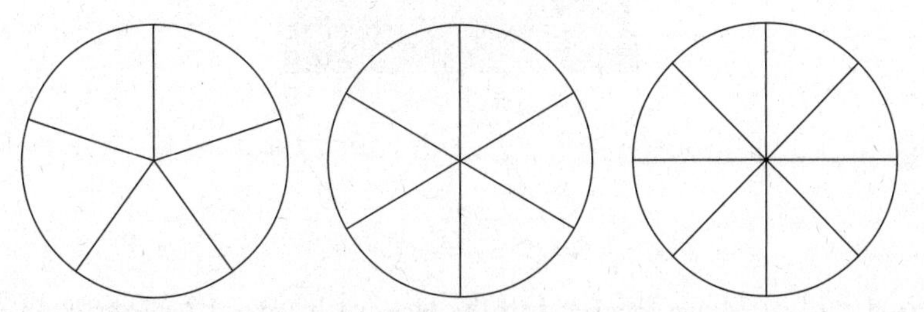

Lindsay has a pie, and she has four friends over, so they can each have a fifth of a pie, as in the first circle.

Another friend comes, and now they can each have a sixth of the pie, as in the second circle.

Finally, two more friends come, and they each will have an eighth, as in the third circle.

Now you try it. Tell what part of the following rectangle is shaded:

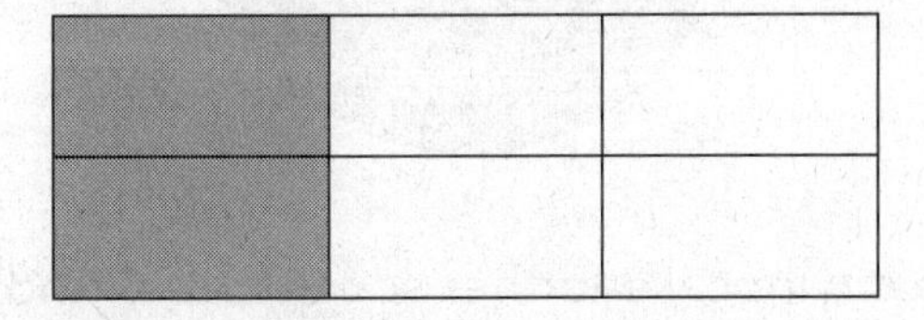

What fraction of the rectangle is shaded? ___________

Did you answer 1/3 of the rectangle? Did you answer 2/6 of the rectangle? Either way, you are right. The same number can be named in many different ways.

Let's try another. Tell what part of this hexagon is shaded:

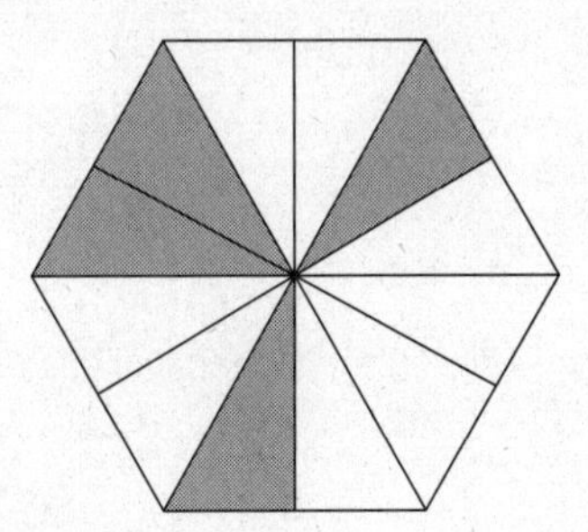

What fraction of the hexagon is shaded? ___________

Did you answer 4/12 of the hexagon? Did you answer 1/3 of the hexagon? Either way, you are right.

Dividing Shapes into Parts Exercises

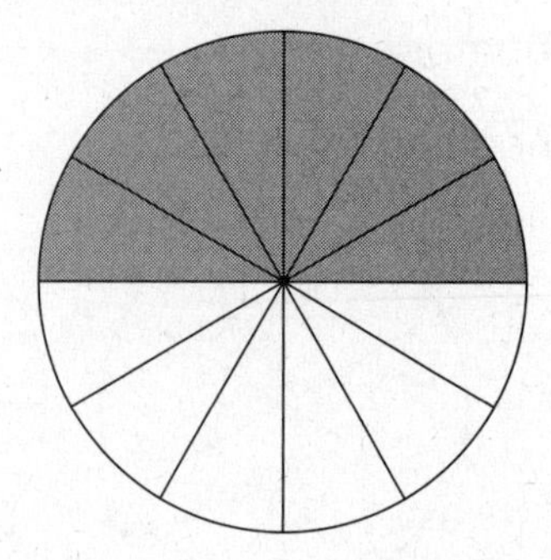

1. Jerry baked a cherry pie for his family. He cut it into 12 pieces, as shown, and his family ate 6 of them. How much was left? Give two ways of writing it.

   There was ______ or ______ of the cherry pie left over.

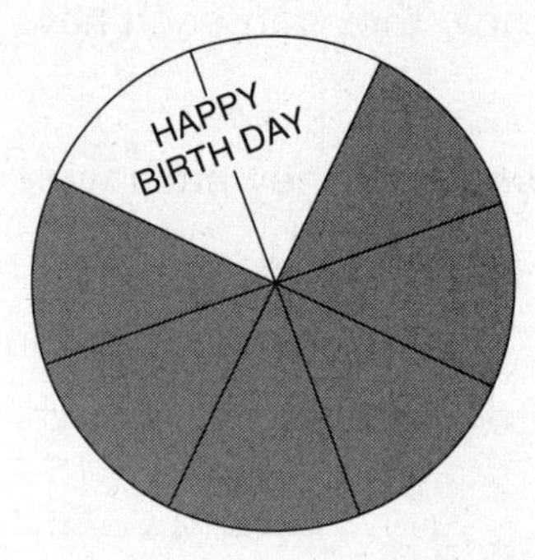

2. Lori ordered an ice cream cake for her son John's birthday. The guests at the party ate 6 of the pieces. How much of the cake did they have left over? Give two ways of writing it.

   They had ______ or ______ of the ice cream cake left over.

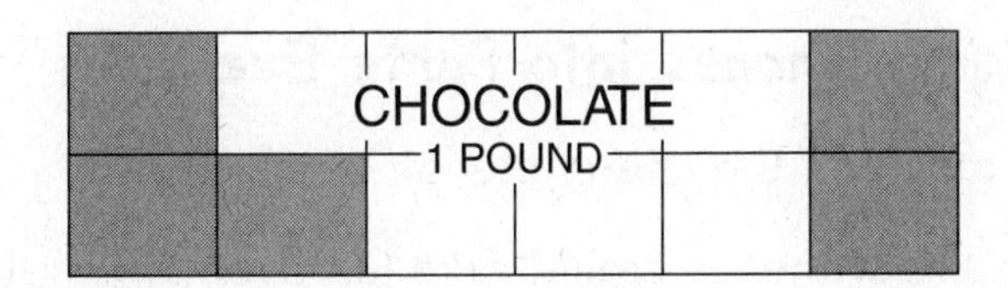

3. Hermi bought a large, one pound chocolate bar. He partitioned it into 12 pieces, as shown, and ate 5 of the pieces (and got sick). How much of the chocolate bar did Hermi eat?

Hermi ate _________ of the chocolate bar.

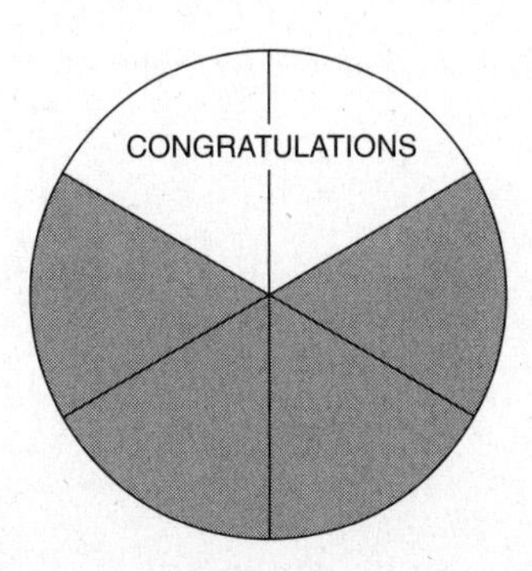

4. For her graduation, Uzma received a large, 12 inch cookie. It was cut into 6 equal pieces, as shown. She and her friends and family ate 4 of the 6 pieces. How much was left over?

There was _______ or _______ of the cookie left over.

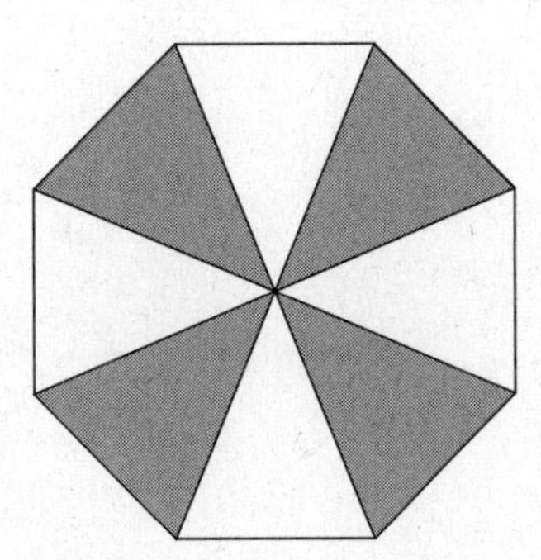

5. Lee makes decorative plates. She made one in the shape of an octagon (having 8 sides). She colored half of the partitions blue, as shown. How much of the plate was white?

The plate was _______ or _______ white.

Answers to Dividing Shapes into Parts Exercises

1. There was 6/12 or 1/2 of the cherry pie left over.
2. They had 2/8 or 1/4 of the ice cream cake left over (which they put in the freezer to avoid melting).
3. Hermi ate 5/12 of the chocolate bar. This cannot be reduced any further.
4. There was 2/6 or 1/3 of the cookie left over.
5. The plate was 4/8 or 1/2 white.

Practice Test—Performance-Based Assessment

CHAPTER 8

Performance-Based Assessment (PBA)

Session 1 (50 minutes)

Directions: This part of the test contains five (5) questions of Type I (4 worth one point each, the fifth worth two points). It also contains three (3) questions of Type II (2 worth three points each, and one worth four points) and one (1) question of Type 3 worth three points. You may NOT use a calculator for these questions. The estimated time for completing this section is 50 minutes; however, you may be permitted more time if you need it.

1. What fraction is greatest? 4/5 1/8 6/12 7/20

Place your answer here ....

2. Roy owns American Royal Hardware, and he received shipments from three suppliers on the first day of the month. From Oliver Tool Co. he received 270 hand tools. From Halstead Garden Supply he received 165 garden tools. From Donovan Paint Co. he received 78 painting tools. How many tools did Roy receive altogether?

Place your answer here ....

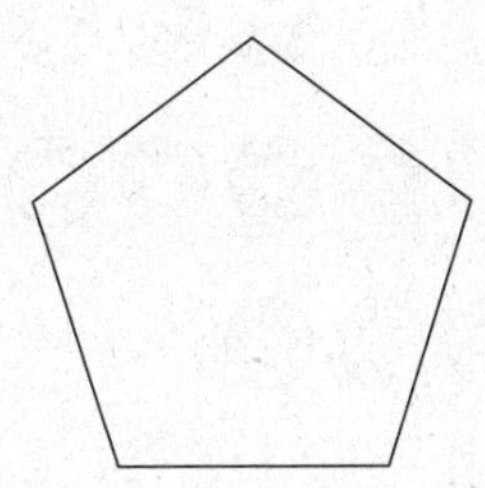

3. What is the name of the shape pictured above?

- ○ A. A square
- ○ B. A hexagon
- ○ C. A triangle
- ○ D. A pentagon

4. Irv keeps track of the customers coming to his car dealership. In the months of summer, he tracked these numbers of customers:

Month	Customers
May	166
June	230
July	351
August	295

How many more customers came to his dealership in July and August than in May and June?

Place your answer here -----------

5. Which two statements could apply to the number fact 4×5?

☐ A. Georgina places 4 rows of 5 cupcakes each on a serving platter
☐ B. Hailey added 4 toy cars to the 5 cars she had
☐ C. In the pool, 5 children join 4 others
☐ D. Frear's Orchard has 4 rows of 5 peach trees
☐ E. Kitera shared 4 cookies with her 5 friends

Place your answer here ----------- -----------

6.

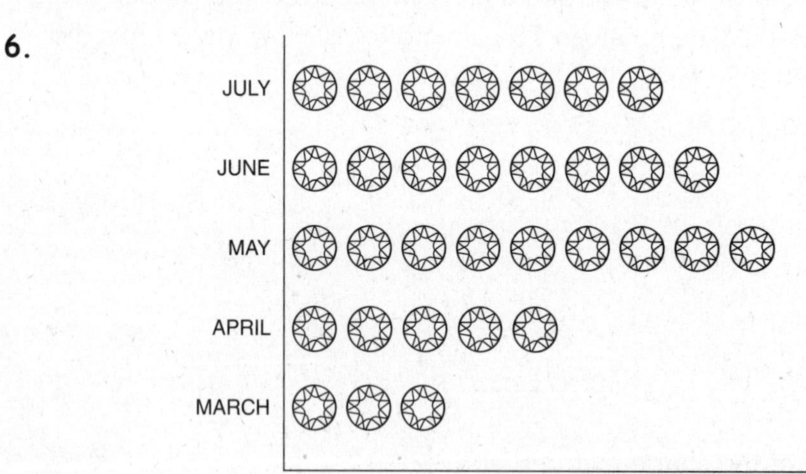

POTHOLES FILLED
(EACH CIRCLE REPRESENTS 10 POTHOLES FILLED)

What three statements are FALSE?

☐ A. The total number of potholes filled in March is 30
☐ B. The number of potholes filled in April is larger than the number of potholes filled in July
☐ C. The number of potholes filled in July is 7
☐ D. The number of potholes filled in April and June is 130
☐ E. There were more potholes filled in March than in May

7. Theo owns Lots 'O Suds Laundromat. Over a 3 week period, his machines laundered:

Week 1: 389 loads
Week 2: 467 loads
Week 3: 292 loads

PART A

How many loads were laundered, in total? ___________

PART B

Theo needs to have 1,500 loads of laundry done every month (every 4 weeks) if he is to stay profitable. How many more loads must he launder in the fourth week, for him to make the 1500 loads?

____

8. Mrs. LaDuca's fourth grade class has a party for her students born in the month of April. She has 8 students, and she has 56 cookies to give the students.

PART A

How many cookies will each student get?

____

Will there be any cookies left, if they do not break the cookies up? How many?

____

PART B

When the students came in, it turned out that two of them were absent. Now, how many cookies will each student get?

____

Will there be any cookies left, if they do not break the cookies up? How many?

____

9. Identify the three figures described below that have an area of 24 square inches.

☐ A.

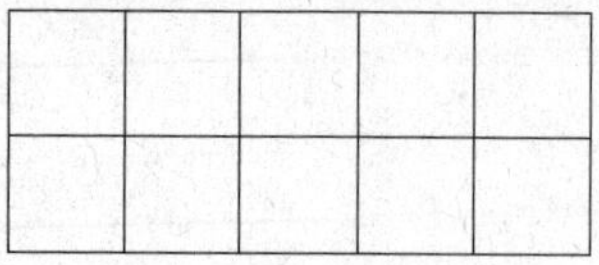

☐ B.

☐ C.

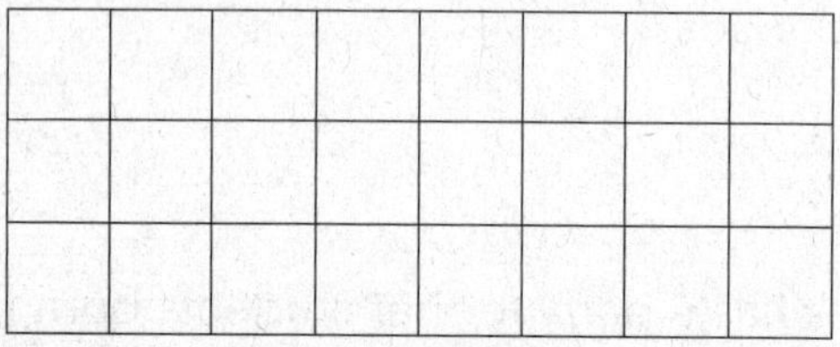

☐ D.

☐ E.

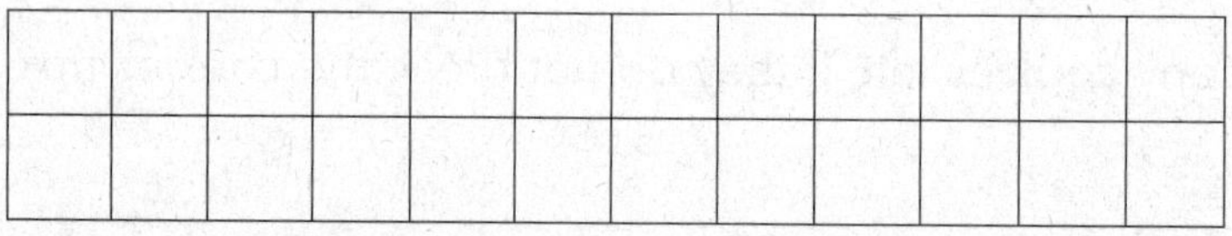

Session 2 (50 minutes)

Directions: This part of the test contains eight (8) questions: four (4) of Type I (worth one point each); one (1) question of Type I (worth two points); one (1) question of Type II (worth four points); one (1) question Type III (worth three points); and one (1) question of Type III (worth 6 points). You may NOT use a calculator for these questions. The estimated time for completing this section is 50 minutes; however, you may be permitted more time if you need it.

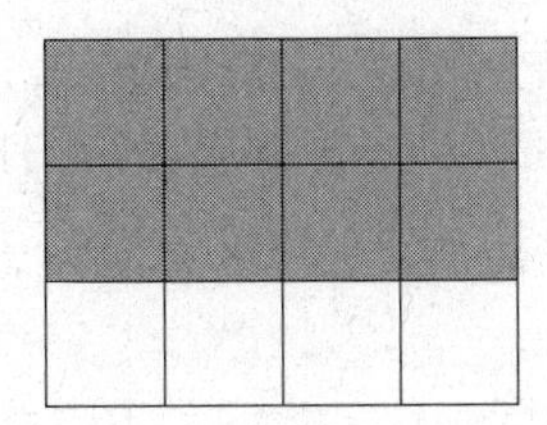

1. What portion of the above figure is *not* colored in?

- ○ A. 1/3
- ○ B. 2/3
- ○ C. 1/2
- ○ D. 4/5

2. Helen works at Circle Cleaners pressing shirts. During one day she pressed:

Hour	8:00	9:00	10:00	11:00
Shirts Pressed	24	28	32	

If this pattern continues, how many shirts will she press at 11:00?

Place your answer here -----------

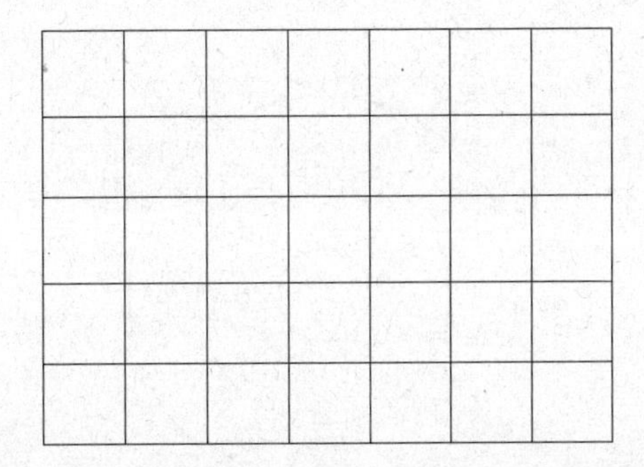

3. Write the multiplication expression that describes the above diagram.

Place your answer here -----------

4. What is 5,473 rounded to the nearest hundred?

Place your answer here -----------

5. Amy went to the store and bought 30 brownies. She bagged them up 4 to a bag for her classmates How many bags of 4 brownies was she able to make? Did she have any brownies left over? How many?

Number of bags of 4 brownies each -----------

Number of brownies left over -----------

6. Flick is stacking boxes in a closet under a staircase. He found that there was a pattern when he counts boxes in the stacks. The first stack has one box, the next has two boxes, the next has three boxes, and the next has four boxes:

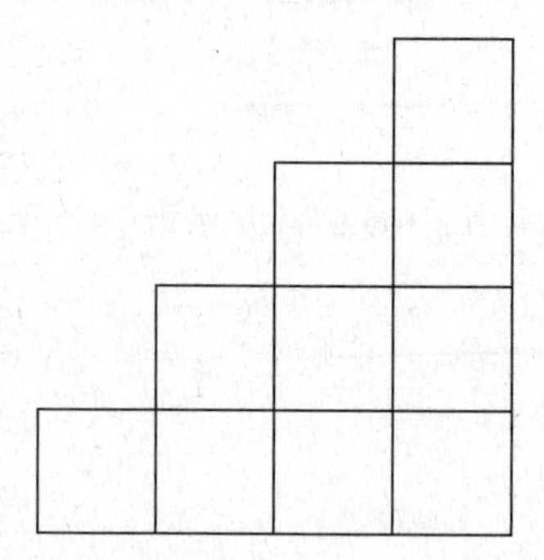

Flick saw a pattern here: The first stack gives one box, the first two stacks give three boxes, the first three stacks give six boxes, and the first four stacks give ten boxes. If this pattern is continued, what will the next three sets of boxes give?

First five stacks ___________

First six stacks ___________

First seven stacks ___________

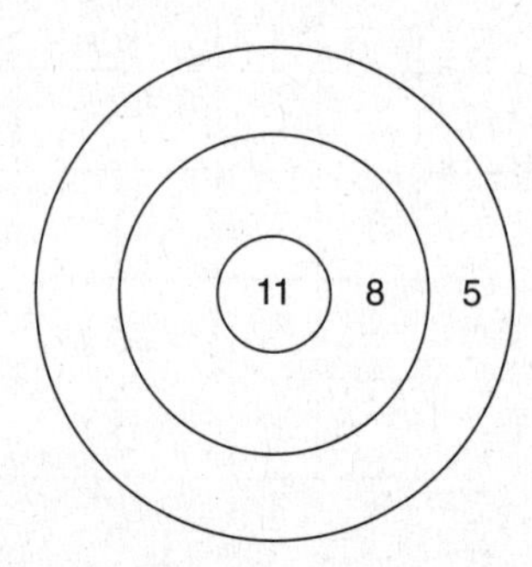

7. Ahmal is playing a dart game using the target above. Each play is with 4 darts. The inner ring is worth 11 points. The middle ring is worth 8 points. The outer ring is worth 5 points. If the dart misses the target, the player loses 4 points.

On his first play, Ahmal hits the inner ring twice, hits the middle ring once, and misses the target with the fourth dart.

What is his score? ___________

On his second play, Ahmal hits the inner ring once, the middle ring once, and the outer ring twice.

What is his score? ___________

On his third play, Ahmal hits the inner ring once, middle ring once, and the outer ring once and misses the target with the fourth dart.

What is his score? ___________

What is Ahmal's final score for the three plays?

Place your answer here ___________

8. Mike decorates the backstop of the town baseball field with ads from local merchants. The backstop is 12 feet by 15 feet. Each cell is one square foot. He has ads from five different merchants, and they pay for different sized ads.

Andy's Shoes paid for a 3×5 foot ad.

Betty's Bagels paid for a 5×5 foot ad.

Charlie's Convenience Store paid for a 3×4 foot ad.

Dharma's Dry Cleaners paid for a 5×4 foot ad.

Part A

What is the total number of square feet covered with ads?

Place your answer here ___________

Part B

Use the grid below to place the four ads on the available backstop space. Shade in boxes representing the four ads.

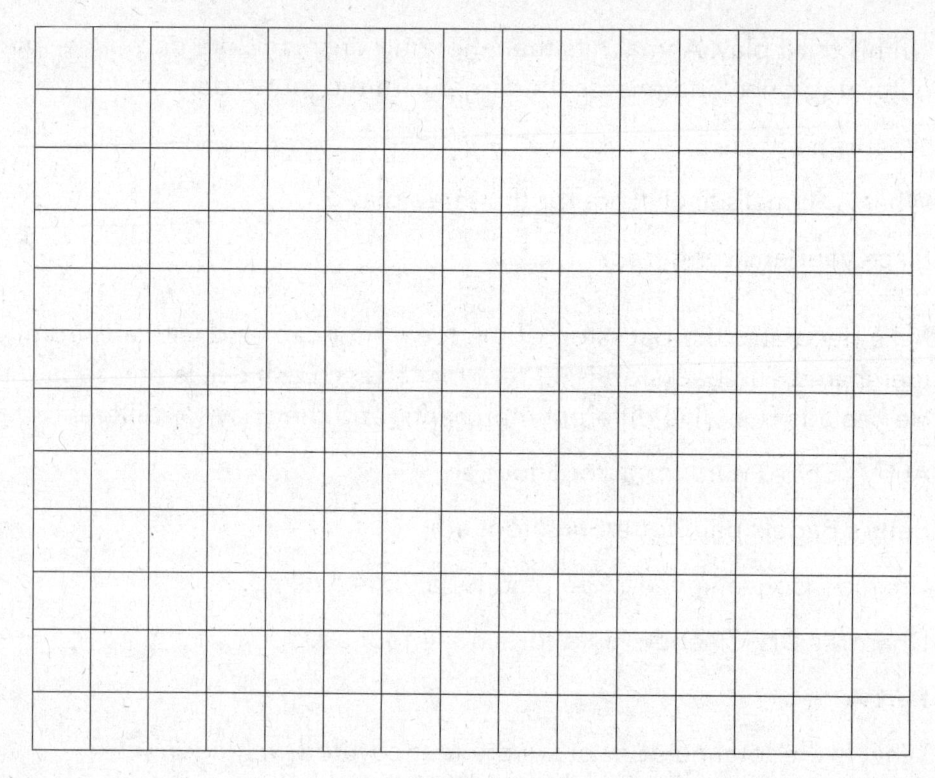

Part C

Mike found that if he arranged the four ads together, he could fit them in a rectangle within the backstop rectangle. What is the difference between the area of the ads and the area of the backstop?

Place your answer here ------------

Answers to Performance-Based Assessment (PBA) Practice Test

Session 1

1. 4/5 is the greatest fraction. 4/5 = 0.8, 1/8 = 0.125, 6/12 = 0.5, 7/20 = 0.35
2. 270 + 165 + 78 = 513 tools
3. **D.** a pentagon
4. 646 - 396 = 250 customers
5. The two statements that are true are:

   A. Georgina places 4 rows of 5 cupcakes each on a serving platter

   D. Frear's Orchard has 4 rows of 5 peach trees

6. The three statements that are false are:

   B. The number of potholes filled in April is larger than the number of potholes filled in July

   C. The number of potholes filled in July is 7

   E. There were more potholes filled in March than in May

7. Part A: There were 1,148 loads of laundry done in the 3 weeks.

   Part B: If Theo needs 1,500 loads laundered in a 4 week period then he needs 1,500 – 1,148 = 352 loads of laundry done in the fourth week.

8. Mrs. LaDuca would give

   Part A: 56 ÷ 8 = 7 cookies per student. There will be no cookies left.

   Part B: 56 ÷ 6 = 9 with a remainder of 2. There will be two cookies left.

9. The three statements that have an area of 24 square inches are:

   B.

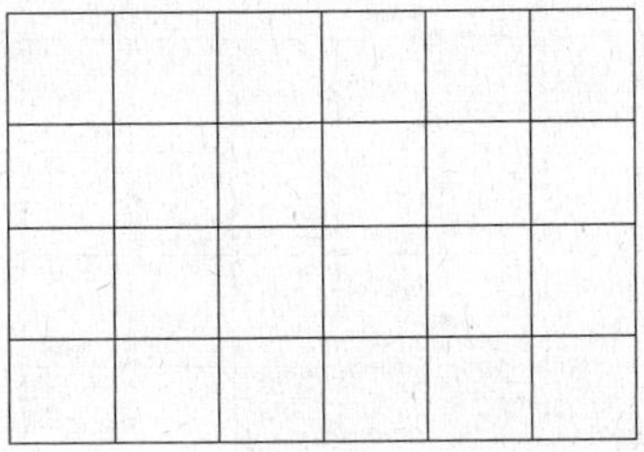

   C.

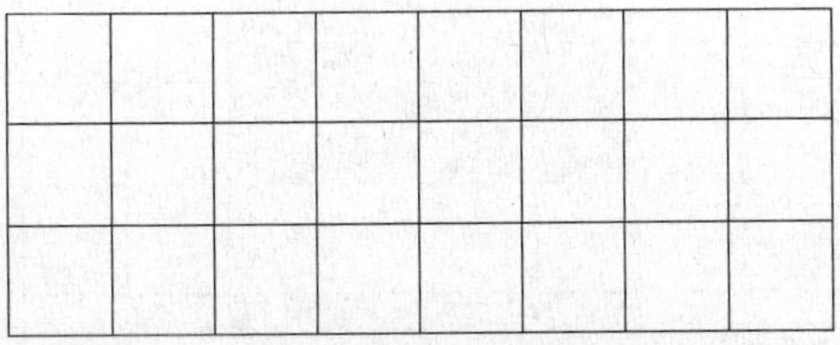

   E.

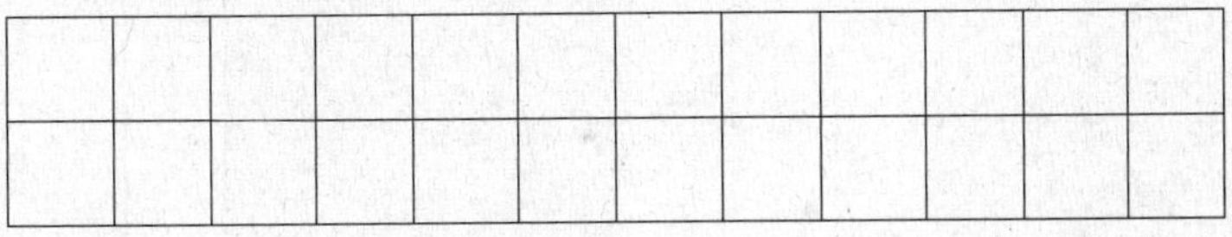

Session 2

1. **A.** 1/3

2. Helen will press 36 shirts at 11:00.

3. 5 × 7 is the expression that describes the figure, but 7 × 5 could also describe the figure.

4. 5,473 rounded to the nearest hundred is 5,500.

5. Amy will make 7 bags of 4 brownies, and there will be 2 brownies left.

6. If the pattern continues, Flick will stack:

   15 boxes with the first five stacks

   21 boxes with the first six stacks

   28 boxes with the first seven stacks

7. On Ahmal's first play, he scores 26 points.

   On Ahmal's second play, he scores 29 points.

   On Ahmal's third play, he scores 20.

   Ahmal's final score is 75.

8. Part A

   Mike finds that the total number of square feet covered with ads is 72 sq ft, which can also be written 72 ft^2.

   Part B

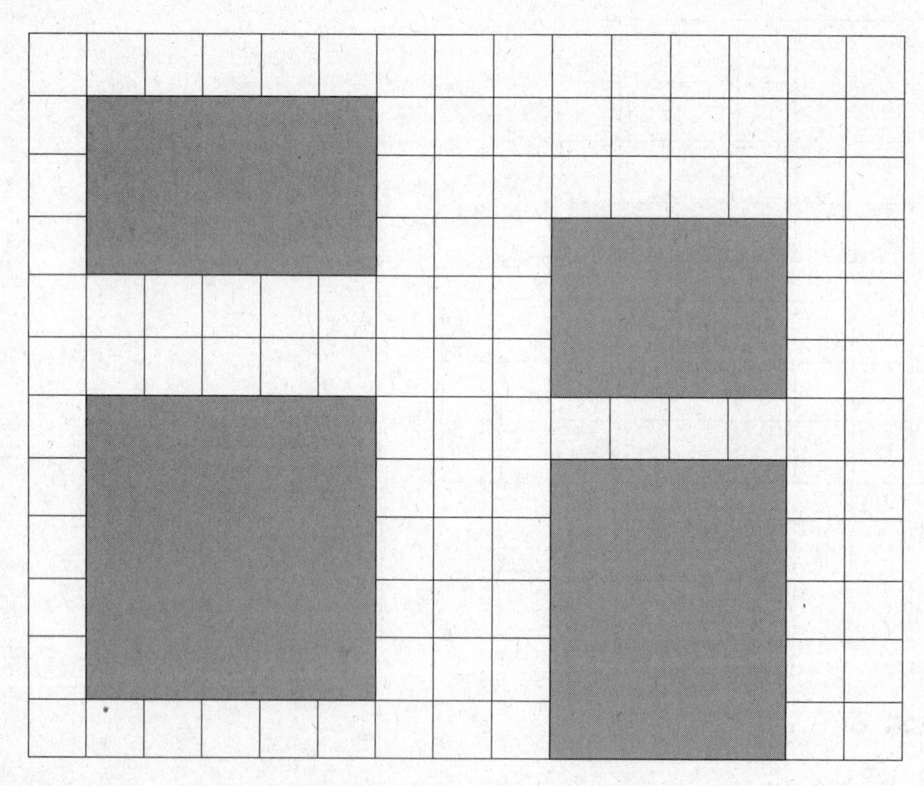

This is one possibility of how to place the ads. There are many ways to arrange the four ads.

Part C

The difference between the area of the ads and the area of the backstop is: 180 − 72 = 108 sq ft, which can also be written as 108 ft^2.

Practice Test— End-of-Year Assessment

CHAPTER 9

End-of-Year Assessment (EOY)

Session 1 (approximately 55 minutes)

Directions: This session of the EOY test contains seventeen (17) Type 1 questions, worth one point each, and two (2) Type 1 questions worth two points each. The estimated time to complete this session is 55 minutes; however, you are permitted to use additional time if necessary. Please complete each question as described.

1. Aiden works at a bagel store. Last week the store made 320 plain bagels, 270 sesame bagels, and 160 whole wheat bagels. How many bagels did the store make last week?

Place your answer here ____________
(unit)

2. What is the area of this rectangle? Each small box is one square centimeter.

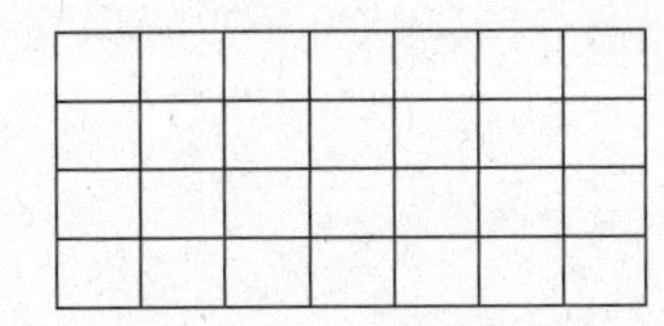

Place your answer here ____________
(unit)

3. What is 687 rounded to the nearest 100?

○ A. 680
○ B. 690
○ C. 600
○ D. 700

4. In the figure below, what portion is <u>not</u> shaded?

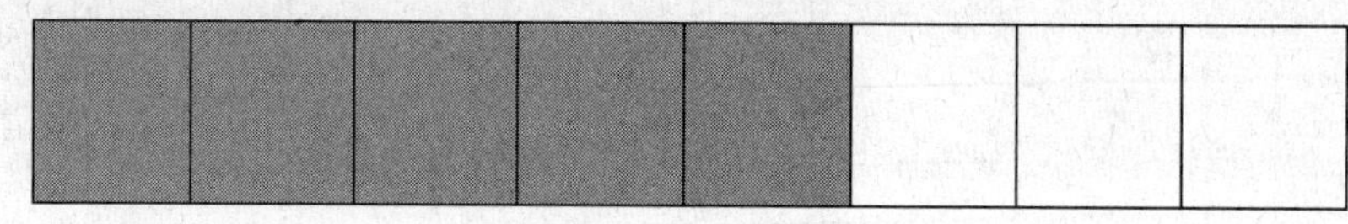

- A. $\frac{5}{8}$
- B. $\frac{4}{8}$
- C. $\frac{3}{8}$
- D. $\frac{2}{8}$

5. In the number line shown:

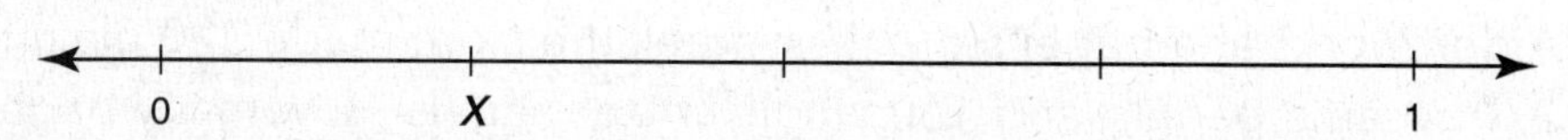

Which fraction would represent the value of the x?

- A. $\frac{4}{1}$
- B. $\frac{1}{1}$
- C. $\frac{1}{3}$
- D. $\frac{1}{4}$

6. About how much water would fill a standard bath tub?

- A. 250 fluid ounces
- B. 250 cups
- C. 250 liters
- D. 250 pints

7. In the following fact triangle, what is the value of t ?

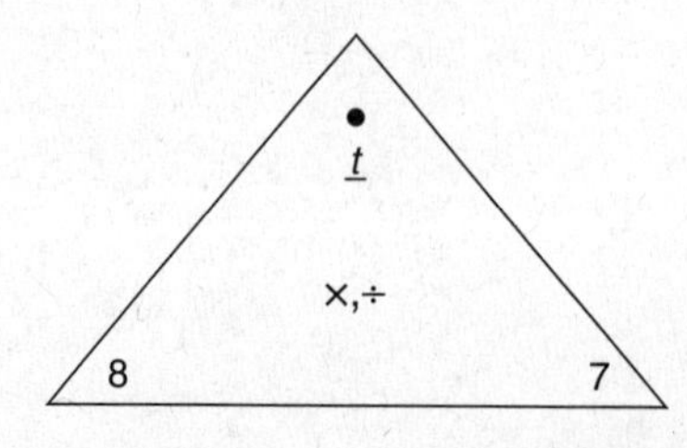

Place your answer here ___________

8. What is the name of the polygon?

Place your answer here ___________

9. What is the area of the shaded region?

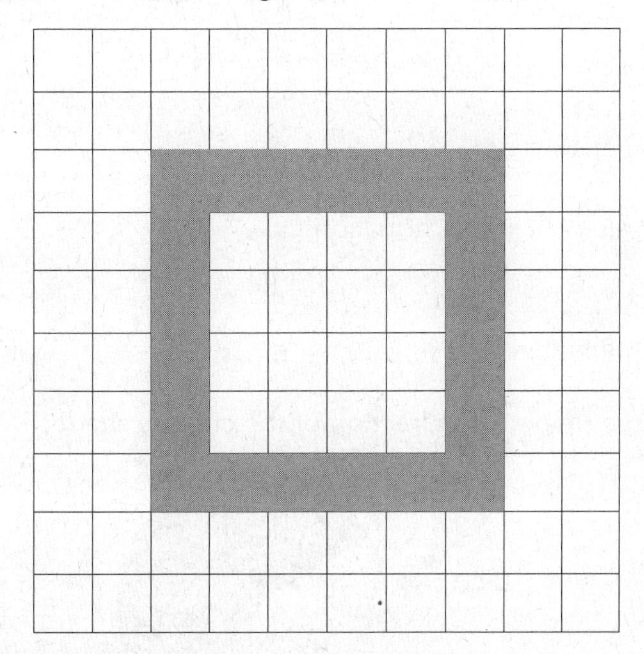

Place your answer here: ___________ square units

10. What time is showing on the clock?

Place your answer here ___________

11. Mr. Gerard has a class trip coming up. He needs to divide up the class so they can ride in cars. He has 24 students in his class, and he has 6 cars to put them in. How many students will go in each car?

Place your answer here ___________
(unit)

12. Write an equivalent fraction to represent the parts that are shaded.

$4 = \frac{?}{?}$

Place your answer here ___________

13. Find the value of r in the equation:

$$r = 84 \div 7$$

Place your answer here ___________

14. In the figure below, what fraction of the circle is shaded?

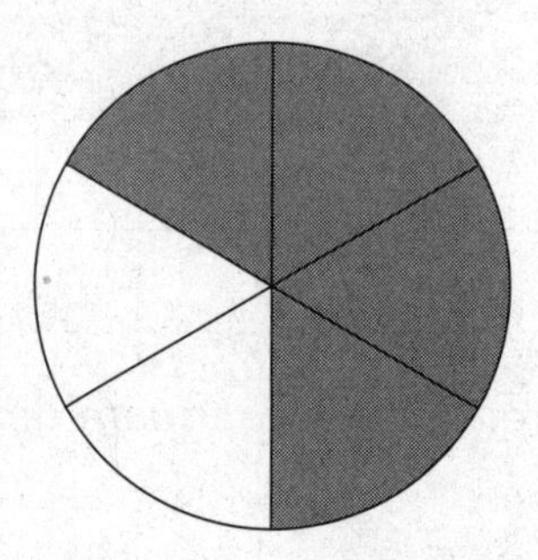

Place your answer here ___________

15. Mrs. Romano took a poll in her class of 24 students to find out the most popular flavor for ice cream. The bar graph depicts her results.

Based on this information, how many more students chose chocolate over strawberry?

○ A. 12 students
○ B. 13 students
○ C. 15 students
○ D. 10 students

16. What is the perimeter of this rectangle? Each square is one square centimeter.

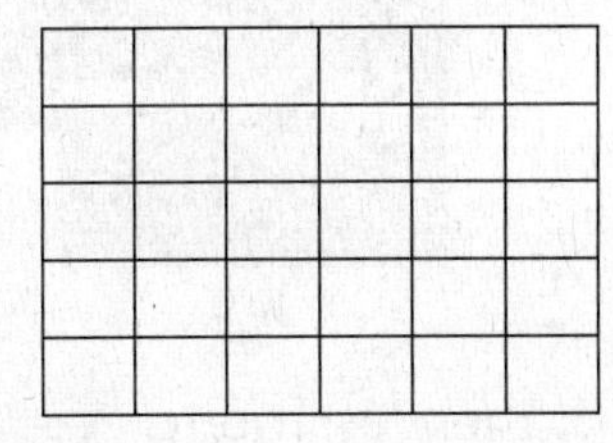

Place your answer here ____________
(unit)

17. Write a number model using fractions and >, <, or = to compare the shaded region of the two bars.

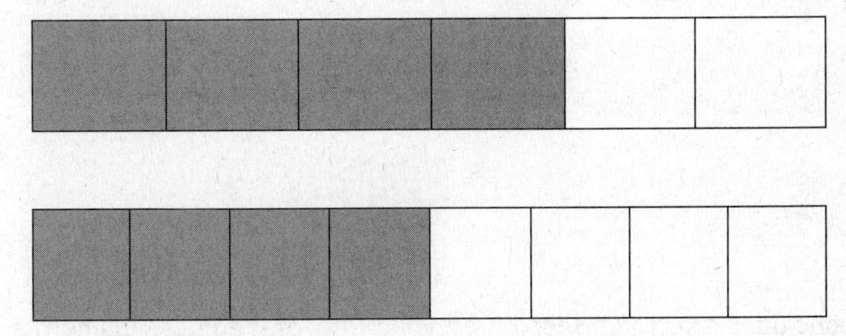

Place your answer here ____________

18. The Wylie's are building a new home. Their design has two windows in the kitchen. One window is 2 ft × 3 ft, and the other window is 3 ft × 3 ft (as shown below).

Window 1:

Window 2:

How many square ft of windows in all?

Place your answer here ___________

19. The picture below is a diagram of a classroom at the Fourth Street Elementary School.

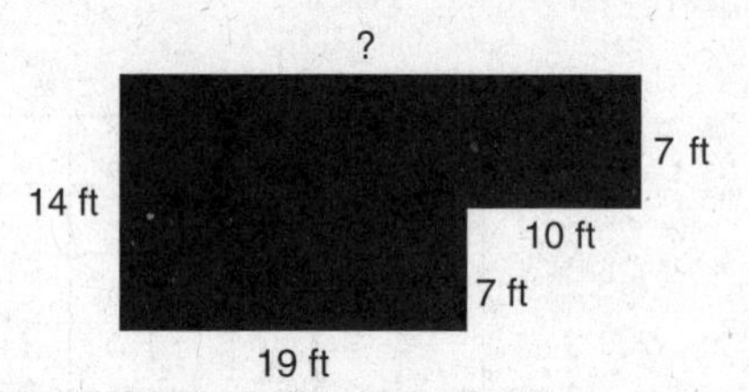

What is the perimeter of the classroom?

Place your answer here: ___________

This concludes the first session of your End-of-Year assessment.

Session 2 (approximately 55 minutes)

Directions: This session of the EOY test contains seventeen (17) Type 1 questions, worth one point each, and three (3) Type 1 questions worth two points each. The estimated time to complete this session is 55 minutes; however, you are permitted to use additional time if necessary. Please complete each question as described.

1. In the following fact triangle, what is the value of *s* ?

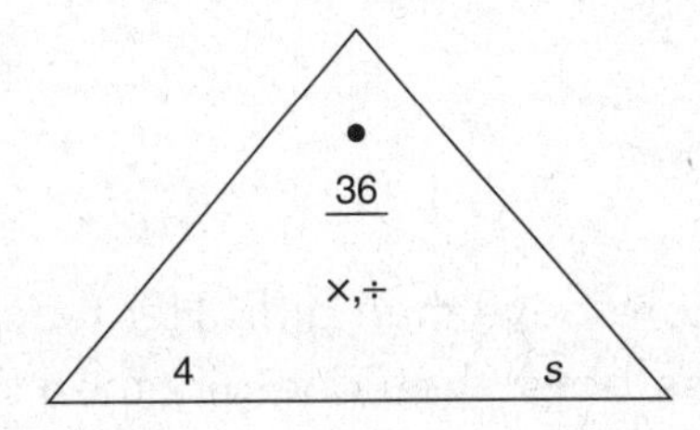

- O A. 9
- O B. 8
- O C. 7
- O D. 6

2. What is the value of *d* in the equation:

$$48 = d \times 6$$

Place your answer here ....

3. What is the total for the shaded area of the bar?

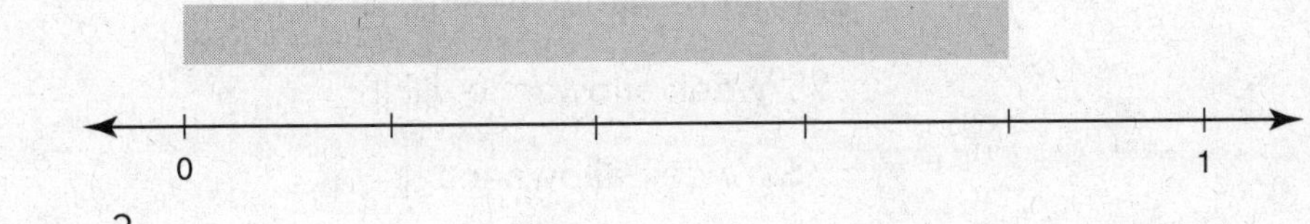

- O A. $\frac{3}{5}$
- O B. $\frac{4}{5}$
- O C. $\frac{2}{5}$
- O D. $\frac{5}{3}$

4. Joaquin's family room measures 20 ft wide by 9 ft long. How many square-foot carpet tiles will Joaquin need to cover the floor?

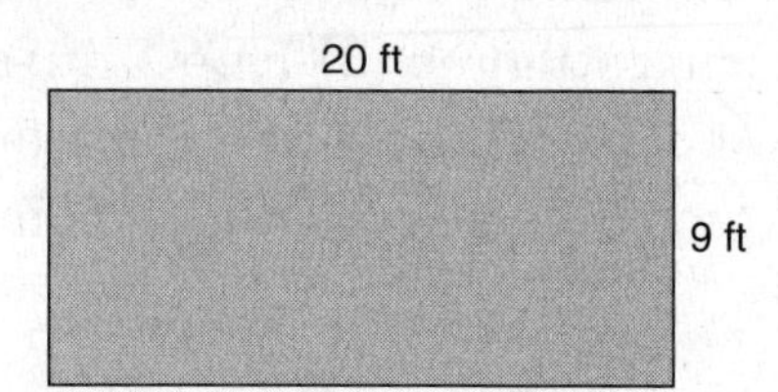

- ○ A. 18 sq ft
- ○ B. 108 sq ft
- ○ C. 11 sq ft
- ○ D. 180 sq ft

5. Hillary's class has more boys than girls. Her teacher lined the students up and was able to put three boys, then one girl, then three boys, then one girl, until the whole class was lined up. If there were 28 students in the class, how many boys were in the class?

Place your answer here ____________
(unit)

6. Mrs. Trossbach's class got into a discussion about doing chores to help out around the house. Many students noted that they get a weekly allowance for doing such chores. Mrs. Trossbach took a poll in her classroom to include everyone's results. Her tallied findings are listed below:

$20/week allowance: 𝍸 |

$10/week allowance: 𝍸 ||||

$5/week allowance: 𝍸 ||

$2/week allowance: |

No allowance: |||

Mrs. Trossbach used this information to graph the chart below.

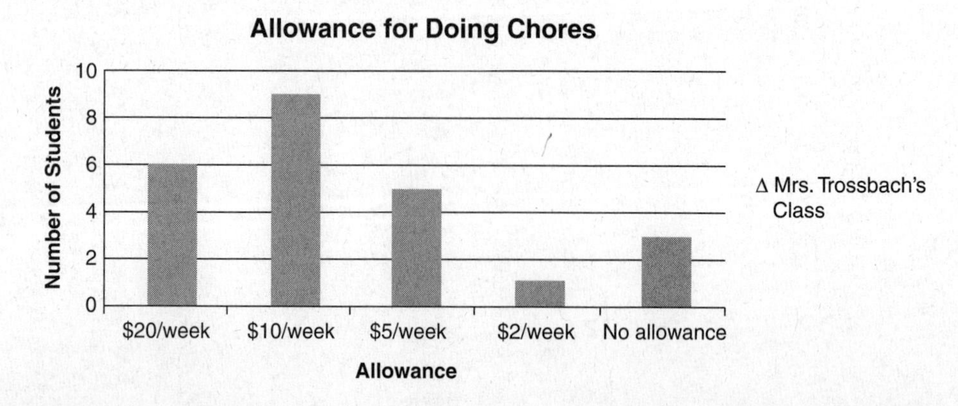

What mistake did Mrs. Trossbach make when she graphed the data?

- ○ A. She shaded the $10/week bar too high
- ○ B. She shaded the $20/week bar too low
- ○ C. She shaded the $5/week bar too low
- ○ D. She shaded the $2/week bar too high

7. Mike and Maya went out to the golf range to practice hitting golf balls. Each bucket had 30 golf balls in it. If they got 9 buckets between the two of them and they hit them all at the range, how many balls did they hit?

Place your answer here -----------
(unit)

8. What is the length of the key to the nearest quarter inch?

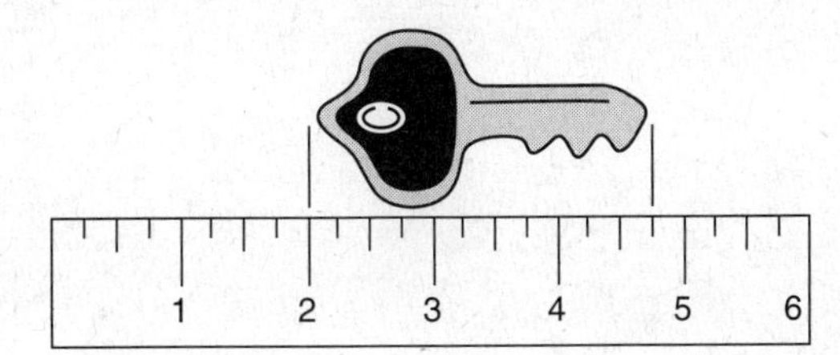

- ○ A. 5 inches
- ○ B. $4\frac{3}{4}$ inches
- ○ C. $4\frac{4}{4}$ inches
- ○ D. $2\frac{3}{4}$ inches

9. Samantha and Ginger sold Girl Scout cookies in front of their local super market. They set a goal to sell 1,000 boxes. So far Samantha sold 350 boxes, and Ginger sold 580 boxes. How many more boxes must be sold to reach their goal?

Place your answer here -----------
(unit)

10. Find the exact answer: 180 ÷ 3

Place your answer here -----------

11. David was walking along the seashore collecting seashells. He collected 30 shells. He then wanted to arrange them in a rectangular pattern. Which array shows David's collection in a rectangular pattern?

○ A.

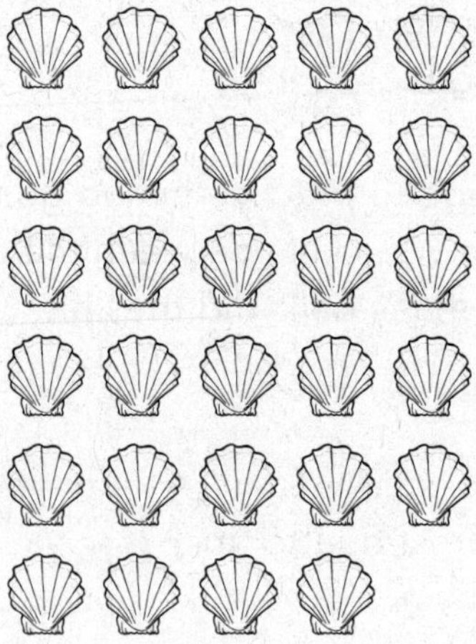

○ B.

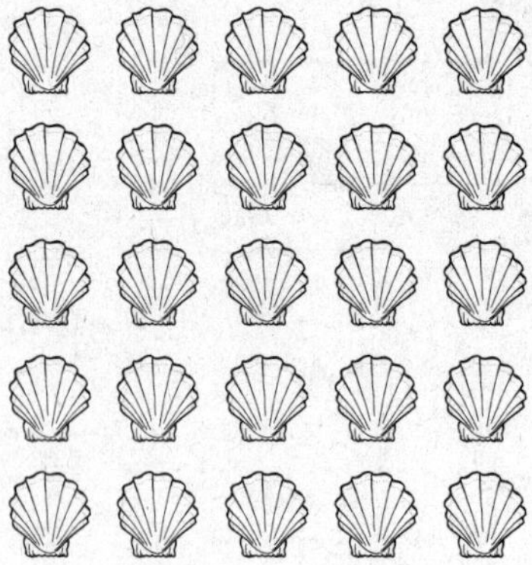

○ C.

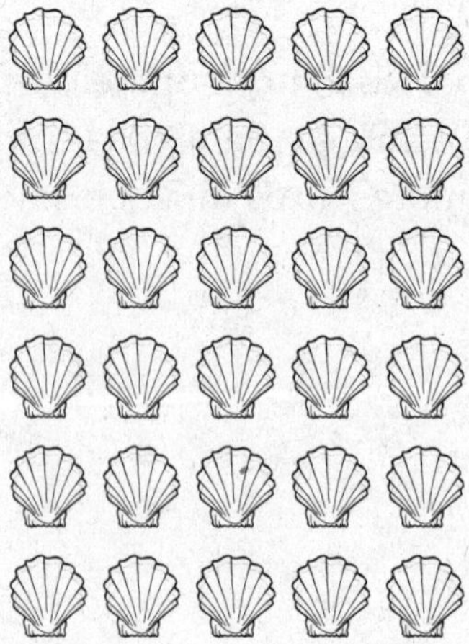

12. Write a number model using >, <, or = to compare $\frac{3}{4}$ and $\frac{3}{8}$.

Place your answer here __________

13. Fatima runs the Bake 'n' More store near Hilltop High School. Last week, she sold cupcakes at the following rate:

Day	Monday	Tuesday	Wednesday	Thursday	Friday
Cupcakes Sold	13	26	39	52	?

If this pattern continues, how many cupcakes will Fatima sell on Friday?

Place your answer here ____________ (unit)

14. Dianna opened up her lunch and saw that she had 5 cookies, each having 8 chocolate chips in them. How many chocolate chips were in the cookies, in all?

- O A. 40 chips
- O B. 48 chips
- O C. 46 chips
- O D. 38 chips

15. Jennifer had a collection of hockey pucks from the games she attended. Before the season started, she had 59 pucks. At the end of the season, she had 71 pucks. How many more pucks did she get that season?

- O A. 130 pucks
- O B. 120 pucks
- O C. 22 pucks
- O D. 12 pucks

16. In the following fact triangle, what is the value of h ?

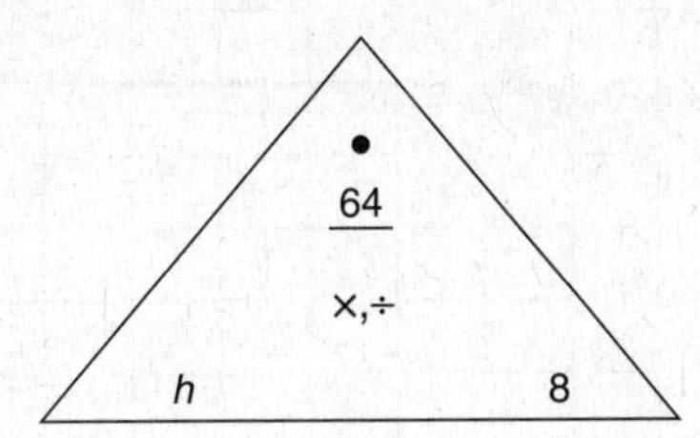

- O A. 5
- O B. 8
- O C. 7
- O D. 6

17. Jim was collecting food for his troop's food drive. He went to 12 houses, and collected the following amounts of food:

House	Cans of Food Collected
1	7
2	5
3	11
4	2
5	14
6	6
7	7
8	6
9	9
10	12
11	15
12	17

How many cans of food did Jim collect for his troop's food drive?

Place your answer here ___________
(unit)

18. Determine the area of the shape:

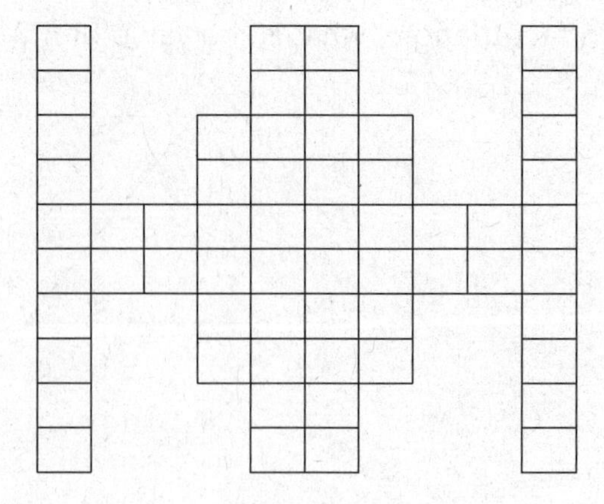

Place your answer here: ___________ square units

19. Determine the area of the shaded region.

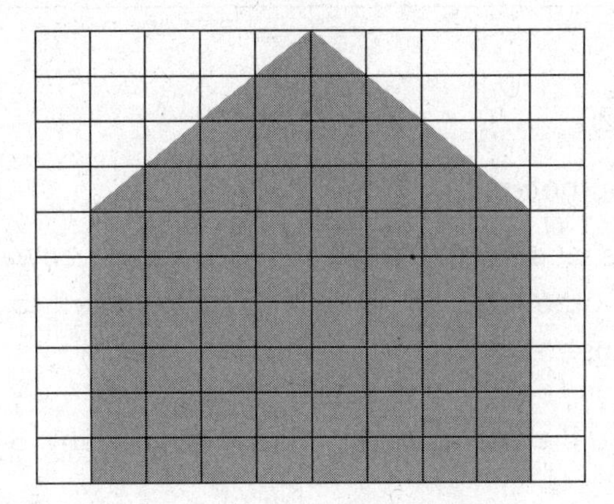

Place your answer here: .... square units

20. The first floor of an apartment is pictured below.

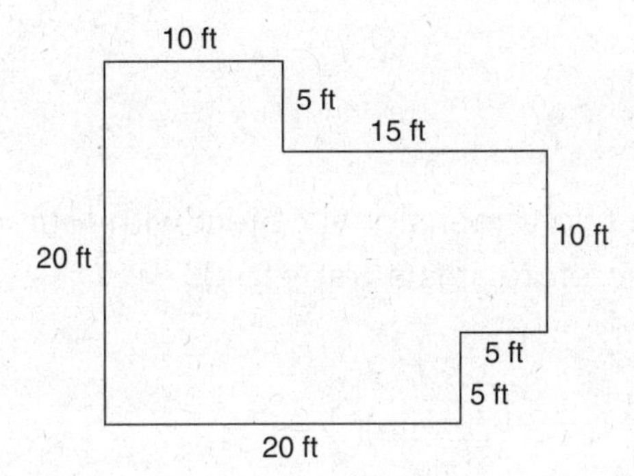

What is the area?

Place your answer here: .... square ft

This concludes the final session of your End-of-Year assessment.

Answers for End-of-Year (EOY) Practice Test

Session 1

1. 750 bagels (320 + 270 + 160 = 750)
2. 28 square centimeters or 28 cm^2
3. **D.** 700; 687 is closer to 700 because it's only 13 away from it. It's 87 away from 600. The other two answers are not rounded to hundreds, they're rounded to tens.
4. **C.** $\frac{3}{8}$; there are 8 equal parts in this rectangle, and 3 of them are not shaded.
5. **D.** $\frac{1}{4}$; there are 4 equal parts on the number line. The *x* shows the endpoint for 1 of the 4 parts.
6. **C.** 250 liters
7. $t = 56$ ($8 \times 7 = 56$)
8. Trapezoid; it is a two-dimensional object with four sides. It also has one set of parallel sides; therefore, it is a trapezoid.
9. 20 square units
10. 2:43 (A.M. or P.M. is acceptable)
11. 4 students ($24 \div 6 = 4$)
12. $4 = \frac{4}{1}$; there are four circles, each divided into only 1 part. Therefore, the denominator of the fraction is 1. There are 4 of them altogether, so the numerator is 4.
13. $r = 12$; one way to solve for r is $84 \div 7 = 12$. Alternately, you could think of it as $r \times 7 = 84$.
14. $\frac{4}{6}$; the circle is divided into 6 equal parts. 4 of those parts are shaded. (It could be reduced to $\frac{2}{3}$.)
15. **B.** 13 students; 15 students chose chocolate, and 2 students chose strawberry. 15 − 2 = 13.
16. 22 centimeters; 5 + 5 + 6 + 6 = 22.

17. $\frac{4}{6} > \frac{4}{8}$

18. 15 sq ft or 15 ft^2 ($2 \times 3 = 6$ and $3 \times 3 = 9$; $6 + 9 = 15$)

19. 86 ft; there is one measure missing from the picture. We can obtain this by adding the two pieces shown: 10 ft + 19 ft = 29 ft.

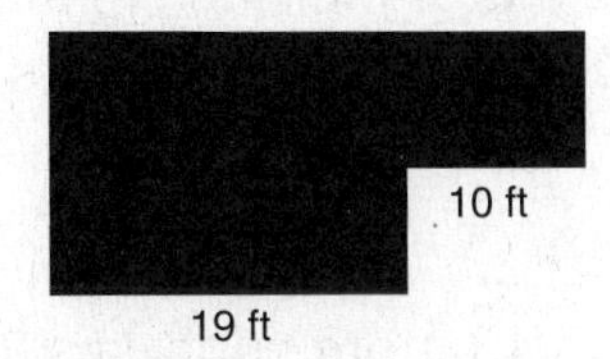

Now that we have all of the measurements, we simply have to add them together: 14 + 19 + 7 + 10 + 7 + 29 = 86 ft.

Session 2

1. **A.** 9 ($36 \div 4 = 9$)
2. $d = 8$; remember, interpret this as 48 is equal to something $\times$ 6.
3. **B.** $\frac{4}{5}$ (The shaded horizontal bar extends over to the endpoint of the fourth part. There are five parts total from 0 to 1, so the fraction the bar represents is $\frac{4}{5}$).
4. **D.** 180; 20 ft $\times$ 9 ft = 180 ft^2. If each tile is a square foot, that means Joaquin will need 180 carpet tiles to cover the floor.
5. 21 boys; the pattern is 3 boys, 1 girl. This is 4 students. There are 7 sets of 4 students because $28 \div 4 = 7$. If there are 7 sets of students, each with 3 boys, then that equals 21 boys altogether because $7 \times 3 = 21$.
6. **C.** She shaded the $5/week bar too low. Mrs. Trossbach tallied 7 students who earned a $5/week allowance; however, she only shaded the bar in to represent 5. Therefore, she shaded the $5/week bar too low.
7. 270 balls; 30 is the same as 3 tens. 9 times 3 tens is 27 tens. 27 tens is 270.
8. **D.** $2\frac{3}{4}$ inches
9. 70 boxes; 350 + 580 = 930 boxes sold. If the goal is 1,000 boxes, then 1,000 – 930 = 70 boxes.
10. $180 \div 3 = 60$ (180 is 18 tens. 18 tens divided by 3 equals 6 tens. 6 tens equals 60).

11. **C.** This is the only correct answer. The array is a rectangle, and it contains all 30 shells in David's collection.

12. $\frac{3}{4} > \frac{3}{8}$; in this equation we have like numerators. Because the numerators are the same, we focus on the denominator. The higher the denominator, the smaller each part is. Since 8 is higher than 4, this tells us the pieces in that fraction are smaller. Since both fractions refer to 3 pieces, we now know that $\frac{3}{8}$ is smaller.

13. 65 cupcakes. The pattern is 13 more cupcakes than the day before: $13 + 13 = 26$; $26 + 13 = 39$; $39 + 13 = 52$; $52 + 13 = 65$.

14. **A.** 40 chips ($8 \times 5 = 40$)

15. **D.** 12 pucks ($71 - 59 = 12$ pucks)

16. **B.** 8 ($64 \div 8 = 8$)

17. 111 cans of food ($7 + 5 + 11 + 2 + 14 + 6 + 7 + 6 + 9 + 12 + 15 + 17 = 111$) Remember, you can group some of these together to simplify the addition process.

18. 60 square units

19. 64 square units; the rectangle on the bottom has an area of 48 square units. The triangle on top has 8 halves, which equals 4 square units. It also has 12 full square units. $48 + 12 + 4 = 64$ square units.

20. 400 sq ft or 400 ft^2. This polygon is made up of 3 separate rectangles. We can find the area of each and then add them together. The top rectangle has an area of $10 \text{ ft} \times 5 \text{ ft} = 50 \text{ ft}^2$. The middle rectangle has an area of $(15 \text{ ft} + 10 \text{ ft}) \times 10 \text{ ft} = 250 \text{ ft}^2$. The bottom rectangle has an area of $5 \text{ ft} \times 20 \text{ ft} = 100 \text{ ft}^2$. If we add those 3 areas, we come up with $50 \text{ ft}^2 + 250 \text{ ft}^2 + 100 \text{ ft}^2 = 400 \text{ ft}^2$).

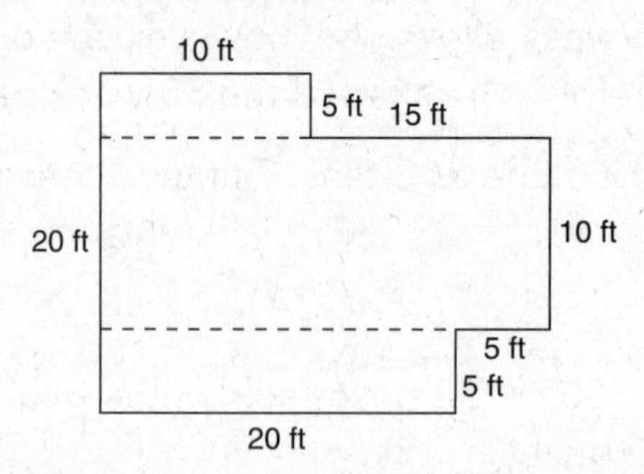

Common Core Standards, Mathematics Grade 3

Domain: Operations & Algebraic Thinking

Cluster: Represent and Solve Problems Involving Multiplication and Division.

Standard: CCSS.MATH.CONTENT.3.OA.A.1

Interpret products of whole numbers, e.g., interpret 5×7 as the total number of objects in 5 groups of 7 objects each. For example, describe a context in which a total number of objects can be expressed as 5×7.

Standard: CCSS.MATH.CONTENT.3.OA.A.2

Interpret whole-number quotients of whole numbers, e.g., interpret $56 \div 8$ as the number of objects in each share when 56 objects are partitioned equally into 8 shares, or as a number of shares when 56 objects are partitioned into equal shares of 8 objects each. For example, describe a context in which a number of shares or a number of groups can be expressed as $56 \div 8$.

Standard: CCSS.MATH.CONTENT.3.OA.A.3

Use multiplication and division within 100 to solve word problems in situations involving equal groups, arrays, and measurement quantities, e.g., by using drawings and equations with a symbol for the unknown number to represent the problem.

Standard: CCSS.MATH.CONTENT.3.OA.A.4

Determine the unknown whole number in a multiplication or division equation relating three whole numbers. For example, determine the unknown number that makes the equation true in each of the equations $8 \times ? = 48$, $5 = \_ \div 3$, $6 \times 6 = ?$

Cluster: Understand Properties of Multiplication and the Relationship between Multiplication and Division.

Standard: CCSS.MATH.CONTENT.3.OA.B.5

Apply properties of operations as strategies to multiply and divide. 2 Examples: If $6 \times 4 = 24$ is known, then $4 \times 6 = 24$ is also known. (Commutative property of multiplication.) $3 \times 5 \times 2$ can be found by $3 \times 5 = 15$, then $15 \times 2 = 30$, or by

$5 \times 2 = 10$, then $3 \times 10 = 30$. (Associative property of multiplication.) Knowing that $8 \times 5 = 40$ and $8 \times 2 = 16$, one can find 8×7 as $8 \times (5 + 2) = (8 \times 5) + (8 \times 2) = 40 + 16 = 56$. (Distributive property.)

Standard: CCSS.MATH.CONTENT.3.OA.B.6

Understand division as an unknown-factor problem. For example, find $32 \div 8$ by finding the number that makes 32 when multiplied by 8.

Cluster: Multiply and Divide within 100.

Standard: CCSS.MATH.CONTENT.3.OA.C.7

Fluently multiply and divide within 100, using strategies such as the relationship between multiplication and division (e.g., knowing that $8 \times 5 = 40$, one knows $40 \div 5 = 8$) or properties of operations. By the end of Grade 3, know from memory all products of two one-digit numbers.

Cluster: Solve Problems Involving the Four Operations, and Identify and Explain Patterns in Arithmetic.

Standard: CCSS.MATH.CONTENT.3.OA.D.8

Solve two-step word problems using the four operations. Represent these problems using equations with a letter standing for the unknown quantity. Assess the reasonableness of answers using mental computation and estimation strategies including rounding.

Standard: CCSS.MATH.CONTENT.3.OA.D.9

Identify arithmetic patterns (including patterns in the addition table or multiplication table), and explain them using properties of operations. For example, observe that 4 times a number is always even, and explain why 4 times a number can be decomposed into two equal addends.

Domain: Number & Operations in Base Ten

Cluster: Use Place Value Understanding and Properties of Operations to Perform Multi-Digit Arithmetic.

Standard: CCSS.MATH.CONTENT.3.NBT.A.1

Use place value understanding to round whole numbers to the nearest 10 or 100.

Standard: CCSS.MATH.CONTENT.3.NBT.A.2

Fluently add and subtract within 1000 using strategies and algorithms based on place value, properties of operations, and/or the relationship between addition and subtraction.

Standard: CCSS.MATH.CONTENT.3.NBT.A.3

Multiply one-digit whole numbers by multiples of 10 in the range 10–90 (e.g., 9×80, 5×60) using strategies based on place value and properties of operations.

Domain: Number & Operations—Fractions

Cluster: Develop Understanding of Fractions as Numbers.

Standard: CCSS.MATH.CONTENT.3.NF.A.1

Understand a fraction $1/b$ as the quantity formed by 1 part when a whole is partitioned into b equal parts; understand a fraction a/b as the quantity formed by a parts of size $1/b$.

Standard: CCSS.MATH.CONTENT.3.NF.A.2

Understand a fraction as a number on the number line; represent fractions on a number line diagram.

Standard: CCSS.MATH.CONTENT.3.NF.A.2.A

Represent a fraction $1/b$ on a number line diagram by defining the interval from 0 to 1 as the whole and partitioning it into b equal parts. Recognize that each part has size $1/b$ and that the endpoint of the part based at 0 locates the number $1/b$ on the number line.

Standard: CCSS.MATH.CONTENT.3.NF.A.2.B

Represent a fraction a/b on a number line diagram by marking off lengths $1/b$ from 0. Recognize that the resulting interval has size a/b and that its endpoint locates the number a/b on the number line.

Standard: CCSS.MATH.CONTENT.3.NF.A.3

Explain equivalence of fractions in special cases, and compare fractions by reasoning about their size.

Standard: CCSS.MATH.CONTENT.3.NF.A.3.A

Understand two fractions as equivalent (equal) if they are the same size, or the same point on a number line.

Standard: CCSS.MATH.CONTENT.3.NF.A.3.B

Recognize and generate simple equivalent fractions, e.g., 1/2 = 2/4, 4/6 = 2/3. Explain why the fractions are equivalent, e.g., by using a visual fraction model.

Standard: CCSS.MATH.CONTENT.3.NF.A.3.C

Express whole numbers as fractions, and recognize fractions that are equivalent to whole numbers. Examples: Express 3 in the form 3 = 3/1; recognize that 6/1 = 6; locate 4/4 and 1 at the same point of a number line diagram.

Standard: CCSS.MATH.CONTENT.3.NF.A.3.D

Compare two fractions with the same numerator or the same denominator by reasoning about their size. Recognize that comparisons are valid only when the two fractions refer to the same whole. Record the results of comparisons with the symbols >, =, or <, and justify the conclusions, e.g., by using a visual fraction model.

Domain: Measurement & Data

Cluster: Solve Problems Involving Measurement and Estimation.

Standard: CCSS.MATH.CONTENT.3.MD.A.1

Tell and write time to the nearest minute and measure time intervals in minutes. Solve word problems involving addition and subtraction of time intervals in minutes, e.g., by representing the problem on a number line diagram.

Standard: CCSS.MATH.CONTENT.3.MD.A.2

Measure and estimate liquid volumes and masses of objects using standard units of grams (g), kilograms (kg), and liters (l). Add, subtract, multiply, or divide to solve one-step word problems involving masses or volumes that are given in the same units, e.g., by using drawings (such as a beaker with a measurement scale) to represent the problem.

Cluster: Represent and Interpret Data.

Standard: CCSS.MATH.CONTENT.3.MD.B.3

Draw a scaled picture graph and a scaled bar graph to represent a data set with several categories. Solve one- and two-step "how many more" and "how many less" problems using information presented in scaled bar graphs. For example, draw a bar graph in which each square in the bar graph might represent 5 pets.

Standard: CCSS.MATH.CONTENT.3.MD.B.4

Generate measurement data by measuring lengths using rulers marked with halves and fourths of an inch. Show the data by making a line plot, where the horizontal scale is marked off in appropriate units–whole numbers, halves, or quarters.

Cluster: Geometric Measurement: Understand Concepts of Area and Relate Area to Multiplication and to Addition.

Standard: CCSS.MATH.CONTENT.3.MD.C.5

Recognize area as an attribute of plane figures and understand concepts of area measurement.

Standard: CCSS.MATH.CONTENT.3.MD.C.5.A

A square with side length 1 unit, called "a unit square," is said to have "one square unit" of area, and can be used to measure area.

Standard: CCSS.MATH.CONTENT.3.MD.C.5.B

A plane figure which can be covered without gaps or overlaps by n unit squares is said to have an area of n square units.

Standard: CCSS.MATH.CONTENT.3.MD.C.6

Measure areas by counting unit squares (square cm, square m, square in, square ft, and improvised units).

Standard: CCSS.MATH.CONTENT.3.MD.C.7

Relate area to the operations of multiplication and addition.

Standard: CCSS.MATH.CONTENT.3.MD.C.7.A

Find the area of a rectangle with whole-number side lengths by tiling it, and show that the area is the same as would be found by multiplying the side lengths.

Standard: CCSS.MATH.CONTENT.3.MD.C.7.B

Multiply side lengths to find areas of rectangles with whole-number side lengths in the context of solving real world and mathematical problems, and represent whole-number products as rectangular areas in mathematical reasoning.

Standard: CCSS.MATH.CONTENT.3.MD.C.7.C

Use tiling to show in a concrete case that the area of a rectangle with whole-number side lengths a and $b + c$ is the sum of $a \times b$ and $a \times c$. Use area models to represent the distributive property in mathematical reasoning.

Standard: CCSS.MATH.CONTENT.3.MD.C.7.D

Recognize area as additive. Find areas of rectilinear figures by decomposing them into non-overlapping rectangles and adding the areas of the non-overlapping parts, applying this technique to solve real world problems.

Cluster: Geometric Measurement: Recognize Perimeter.

Standard: CCSS.MATH.CONTENT.3.MD.D.8

Solve real world and mathematical problems involving perimeters of polygons, including finding the perimeter given the side lengths, finding an unknown side length, and exhibiting rectangles with the same perimeter and different areas or with the same area and different perimeters.

Domain: Measurement & Data

Cluster: Reason with Shapes and Their Attributes.

Standard: CCSS.MATH.CONTENT.3.G.A.1

Understand that shapes in different categories (e.g., rhombuses, rectangles, and others) may share attributes (e.g., having four sides), and that the shared attributes can define a larger category (e.g., quadrilaterals). Recognize rhombuses, rectangles, and squares as examples of quadrilaterals, and draw examples of quadrilaterals that do not belong to any of these subcategories.

Standard: CCSS.MATH.CONTENT.3.G.A.2

Partition shapes into parts with equal areas. Express the area of each part as a unit fraction of the whole. For example, partition a shape into 4 parts with equal area, and describe the area of each part as 1/4 of the area of the shape.

Standards for Mathematical Practice

APPENDIX B

The Standards for Mathematical Practice describe varieties of expertise that mathematics educators at all levels should seek to develop in their students. These practices rest on important "processes and proficiencies" with longstanding importance in mathematics education. The first of these are the NCTM process standards of problem solving, reasoning and proof, communication, representation, and connections. The second are the strands of mathematical proficiency specified in the National Research Council's report Adding It Up : adaptive reasoning, strategic competence, conceptual understanding (comprehension of mathematical concepts, operations and relations), procedural fluency (skill in carrying out procedures flexibly, accurately, efficiently and appropriately), and productive disposition (habitual inclination to see mathematics as sensible, useful, and worthwhile, coupled with a belief in diligence and one's own efficacy).

Standards in the Mathematical Practice Domains

CCSS.MATH.PRACTICE.MP1

Make sense of problems and persevere in solving them.

Mathematically proficient students start by explaining to themselves the meaning of a problem and looking for entry points to its solution. They analyze givens, constraints, relationships, and goals. They make conjectures about the form and meaning of the solution and plan a solution pathway rather than simply jumping into a solution attempt. They consider analogous problems, and try special cases and simpler forms of the original problem in order to gain insight into its solution. They monitor and evaluate their progress and change course if necessary. Older students might, depending on the context of the problem, transform algebraic expressions or change the viewing window on their graphing calculator to get the information they need. Mathematically proficient students can explain correspondences between equations, verbal descriptions, tables, and graphs or draw diagrams of important features and relationships, graph data, and search for regularity or trends. Younger students might rely on using concrete objects or pictures to help conceptualize and solve a problem. Mathematically proficient students check their answers to problems using a different method, and they continually ask themselves, "Does this make sense?" They can understand the approaches of others to solving complex problems and identify correspondences between different approaches.

CCSS.MATH.PRACTICE.MP2

Reason abstractly and quantitatively. Mathematically proficient students make sense of quantities and their relationships in problem situations. They bring two complementary abilities to bear on problems involving quantitative relationships: the ability to decontextualize–to abstract a given situation and represent it symbolically and manipulate the representing symbols as if they have a life of their own, without necessarily attending to their referents–and the ability to contextualize, to pause as needed during the manipulation process in order to probe into the referents for the symbols involved. Quantitative reasoning entails habits of creating a coherent representation of the problem at hand; considering the units involved; attending to the meaning of quantities, not just how to compute them; and knowing and flexibly using different properties of operations and objects.

CCSS.MATH.PRACTICE.MP3

Construct viable arguments and critique the reasoning of others. Mathematically proficient students understand and use stated assumptions, definitions, and previously established results in constructing arguments. They make conjectures and build a logical progression of statements to explore the truth of their conjectures. They are able to analyze situations by breaking them into cases, and can recognize and use counterexamples. They justify their conclusions, communicate them to others, and respond to the arguments of others. They reason inductively about data, making plausible arguments that take into account the context from which the data arose. Mathematically proficient students are also able to compare the effectiveness of two plausible arguments, distinguish correct logic or reasoning from that which is flawed, and–if there is a flaw in an argument–explain what it is. Elementary students can construct arguments using concrete referents such as objects, drawings, diagrams, and actions. Such arguments can make sense and be correct, even though they are not generalized or made formal until later grades. Later, students learn to determine domains to which an argument applies. Students at all grades can listen or read the arguments of others, decide whether they make sense, and ask useful questions to clarify or improve the arguments.

CCSS.MATH.PRACTICE.MP4

Model with mathematics. Mathematically proficient students can apply the mathematics they know to solve problems arising in everyday life, society, and the workplace. In early grades, this might be as simple as writing an addition equation to describe a situation. In middle grades, a student might apply proportional reasoning to plan a school event or analyze a problem in the community. By high school, a student might use geometry to solve a design problem or use a function to describe how one quantity of interest depends on another. Mathematically proficient students who can apply what they know are comfortable making assumptions and approximations to simplify a complicated situation, realizing that these may need revision later. They are able to identify important quantities in a practical situation and map their relationships using such tools as diagrams, two-way tables, graphs, flowcharts, and formulas. They can analyze those relationships mathematically to draw conclusions. They routinely interpret their mathematical results in the context of the situation and reflect on whether the results make sense, possibly improving the model if it has not served its purpose.

CCSS.MATH.PRACTICE.MP5

Use appropriate tools strategically. Mathematically proficient students consider the available tools when solving a mathematical problem. These tools might include pencil and paper, concrete models, a ruler, a protractor, a calculator, a spreadsheet, a computer algebra system, a statistical package, or dynamic geometry software. Proficient students are sufficiently familiar with tools appropriate for their grade or course to make sound decisions about when each of these tools might be helpful, recognizing both the insight to be gained and their limitations. For example, mathematically proficient high school students analyze graphs of functions and solutions generated using a graphing calculator. They detect possible errors by strategically using estimation and other mathematical knowledge. When making mathematical models, they know that technology can enable them to visualize the results of varying assumptions, explore consequences, and compare predictions with data. Mathematically proficient students at various grade levels are able to identify relevant external mathematical resources, such as digital content located on a website, and use them to pose or solve problems. They are able to use technological tools to explore and deepen their understanding of concepts.

CCSS.MATH.PRACTICE.MP6

Attend to precision. Mathematically proficient students try to communicate precisely to others. They try to use clear definitions in discussion with others and in their own reasoning. They state the meaning of the symbols they choose, including using the equals sign consistently and appropriately. They are careful about specifying units of measure, and labeling axes to clarify the correspondence with quantities in a problem. They calculate accurately and efficiently, express numerical answers with a degree of precision appropriate for the problem context. In the elementary grades, students give carefully formulated explanations to each other. By the time they reach high school, they have learned to examine claims and make explicit use of definitions.

CCSS.MATH.PRACTICE.MP7

Look for and make use of structure. Mathematically proficient students look closely to discern a pattern or structure. Young students, for example, might notice that three and seven more is the same amount as seven and three more, or they may sort a collection of shapes according to how many sides the shapes have. Later, students will see 7×8 equals the well-remembered $7 \times 5 + 7 \times 3$, in preparation for learning about the distributive property. In the expression $x^2 + 9x + 14$, older students can see the 14 as 2×7 and the 9 as $2 + 7$. They recognize the significance of an existing line in a geometric figure and can use the strategy of drawing an auxiliary line for solving problems. They also can step back for an overview and shift perspective. They can see complicated things, such as some algebraic expressions, as single objects or as being composed of several objects. For example, they can see $5 - 3(x - y)^2$ as 5 minus a positive number times a square and use that to realize that its value cannot be more than 5 for any real numbers x and y.

CCSS.MATH.PRACTICE.MP8

Look for and express regularity in repeated reasoning. Mathematically proficient students notice if calculations are repeated, and look both for general methods and for shortcuts. Upper elementary students might notice when dividing 25 by 11 that they are repeating the same calculations over and over again, and conclude they have a repeating decimal. By paying attention to the calculation of slope as they repeatedly check whether points are on the line through (1, 2) with slope 3, middle school students might abstract the equation $(y - 2)/(x - 1) = 3$. Noticing the regularity in the way terms cancel when expanding $(x - 1)(x + 1)$, $(x - 1)(x^2 + x + 1)$, and $(x - 1)(x^3 + x^2 + x + 1)$ might lead them to the general formula for the sum of a geometric series. As they work to solve a problem, mathematically proficient students maintain oversight of the process, while attending to the details. They continually evaluate the reasonableness of their intermediate results.

Connecting the Standards for Mathematical Practice to the Standards for Mathematical Content

The Standards for Mathematical Practice describe ways in which developing student practitioners of the discipline of mathematics increasingly ought to engage with the subject matter as they grow in mathematical maturity and expertise throughout the elementary, middle, and high school years. Designers of curricula, assessments, and professional development should all attend to the need to connect the mathematical practices to mathematical content in mathematics instruction.

The Standards for Mathematical Content are a balanced combination of procedure and understanding. Expectations that begin with the word "understand" are often especially good opportunities to connect the practices to the content. Students who lack understanding of a topic may rely on procedures too heavily. Without a flexible base from which to work, they may be less likely to consider analogous problems, represent problems coherently, justify conclusions, apply the mathematics to practical situations, use technology mindfully to work with the mathematics, explain the mathematics accurately to other students, step back for an overview, or deviate from a known procedure to find a shortcut. In short, a lack of understanding effectively prevents a student from engaging in the mathematical practices.

In this respect, those content standards which set an expectation of understanding are potential "points of intersection" between the Standards for Mathematical Content and the Standards for Mathematical Practice. These points of intersection are intended to be weighted toward central and generative concepts in the school mathematics curriculum that most merit the time, resources, innovative energies, and focus necessary to qualitatively improve the curriculum, instruction, assessment, professional development, and student achievement in mathematics.

Index

NOTES

NOTES

BARRON'S

BARRON'S
COMMON CORE SUCCESS
LEARN, REVIEW, APPLY
GRADE 1 MATH
2+2=
BARRON'S
COMMON CORE SUCCESS
LEARN, REVIEW, APPLY
GRADE 5 LANGUAGE ARTS
what's your STORY
Preparing Students for a Brilliant Future